AN ATLAS TO
Human Anatomy

Dennis Strete
Christopher H. Creek

McGraw-Hill

Boston   Burr Ridge, IL   Dubuque, IA   Madison, WI   New York   San Francisco   St. Louis
Bangkok   Bogotá   Caracas   Lisbon   London   Madrid
Mexico City   Milan   New Delhi   Seoul   Singapore   Sydney   Taipei   Toronto

McGraw-Hill Higher Education

*A Division of The **McGraw-Hill** Companies*

AN ATLAS TO HUMAN ANATOMY

 This book is printed on recycled, acid-free paper containing 10% postconsumer waste.

7 8 9 0 QPD/QPD 0 9 8 7 6 5 4 3

ISBN 0–697–38793–3

Vice president and editorial director: *Kevin T. Kane*
Publisher: *Colin H. Wheatley*
Sponsoring editor: *Kristine Tibbetts*
Developmental editor: *Patrick F. Anglin*
Marketing manager: *Heather K. Wagner*
Senior project manager: *Marilyn Rothenberger*
Senior production supervisor: *Sandra Hahn*
Designer: *K. Wayne Harms*
Supplement coordinator: *Sandra M. Schnee*
Compositor: *Carlisle Communications, Ltd.*
Typeface: *10/12 Times Roman*
Printer: *Quebecor Printing Book Group/Dubuque, IA*

Cover/interior design: *Kristyn Kalnes*

Library of Congress Cataloging-in-Publication Data

Strete, Dennis.
    An atlas to human anatomy / Dennis Strete, Chris Creek. — 1st ed.
       p.   cm.
    Includes index.
    ISBN 0–697–38793–3
    1. Human anatomy—Atlases. I. Creek, Chris. II. Title.
QM25.S767 2000
611′.0022′2—dc21   99–12044
                                    CIP

www.mhhe.com

This atlas is dedicated to my parents:
the late Drs. Alfred and Nellie Strete

Dennis Strete

In dedication to my parents, Dennis and Elna Creek,
and to my wife and children, Lorie, Ryan, Kyle, and Cameron
for their love and support.

Christopher H. Creek

BRIEF CONTENTS

CONTENTS

PREFACE

T his atlas of human anatomy is a result of endless hours of cadaver dissections, photography, script writing, and creation of art work that accompanies many of the images. Such an undertaking would not have been possible without the generosity and encouragement of the anatomy faculty and the administration of Parker College of Chiropractic in Dallas, Texas. Our deep appreciation and thanks go to the following faculty at Parker College: Dr. Steve W. Kirk for his suggestions in organizing the chapter on muscles for this atlas, to Dr. Evans LaFlore for his meticulous cadaver dissections, to Dr. Farshid Marzban for his input in the neuroanatomy section of the atlas, and to Dr. Gene Giggleman for allowing the use of the gross anatomy lab at Parker College.

The authors are also indebted to Kimberly Caplinger for her dedication in typing the manuscript, and to Sandra Gomez for her diligent and conscientious effort in labeling the skeletal system.

Our sincere thanks also go to the following individuals at Scott and White Medical Center, Temple, Texas: Mr. R. Thomas King and Janice Pennington of the Pathology/Electron Microscopy department for contributing some of the electron micrographs; to Vicki Malone of the Anatomical Cytology department for supplying male and female karyotype photographs; and to David S. Henson, Director, Biomedical Communications, and Natalie E. Hubbard for giving their expert opinion in photographing slides and cadavers for the atlas.

Further, the authors appreciate the help of Dr. Robert Allison for establishing the liaison between author Dennis Strete and the Department of Anatomy at Parker College of Chiropractic.

Finally, our special thanks and appreciation go to Kristine Tibbetts and the staff of WCB/McGraw-Hill for publishing this color atlas of human anatomy.

It is our sincere hope that students pursuing anatomy and physiology courses will find this atlas helpful.

Thank you to the following reviewers:

Mary F. Barbe   *Temple University*
Jay Dee Druecker   *Chadron State College*
Leah Dvorak   *Concordia University Wisconsin*
Richard L. Faircloth   *Anne Arundel Community College*
Brian D. Feige   *Mott Community College*
Robert M. George   *Florida International University*
Karen M. LaFleur   *Greenville Technical College*
Corrie A. Mancinelli   *West Virginia University School of Medicine*
Royce Lee Montgomery   *University of North Carolina School of Medicine*
Fontaine C. Piper   *Truman State University*
Michael Jay Shively   *Utah Valley State College*
Curt Walker   *Dixie College*
Charles R. Wert   *Linn-Benton Community College*

Dennis Strete, Ph.D.
Christopher H. Creek

Cells and Tissues

CHAPTER ONE

Anatomy at the Cellular Level

Cells in a given tissue are the basic functional units. In the embryonic stage, cells differentiate into four complex tissues: epithelial tissue, connective tissue, muscular tissue, and nervous tissue. Within a given tissue, a living cell is a mixture of organelles, cytoplasm, nucleus, and membranes. Within the cell, complex excitable units respond to environmental variations and continuously exchange energy and excitable molecules with their immediate surroundings. It is difficult to comprehend how a structure so small, which can be seen only with the help of a microscope, can metabolize, repair itself, go through mitotic and meiotic divisions, and not only create its own daughter cells but groups of cells that differentiate into germinal sperm and ovum cells in later stages of life.

By studying cells under an electron microscope, one finds that the cell membrane is a highly specialized combination of proteins and lipid molecules. The lipid layer in the membrane forms its main internal fabric, whereas the surface protein layers determine specific functions of the membrane. The membranes of the organelles and the plasma membrane of the cell have a unique arrangement of proteins and lipids that facilitate the membranes to be selectively permeable to some molecules and not permeable to others. Functionally, membranes are involved in the transport of small proteins, the diffusion and osmosis of ions and water, the maintaining of electrochemical gradients, the maintaining of endocytosis and exocytosis of various kinds, and the regulation of signal-transduction pathways that enable the intracellular environment of the cell to communicate with the extracellular environment.

The membrane-bound intracellular organelles provide metabolic functions with the help of specific enzymes. The cell nucleus, which controls all cellular reactions, is surrounded by a bilayer nuclear envelope with communicating nuclear pores that connect the internal nuclear nucleoplasm to the external nuclear environment. The membrane-bound ribosomes and polysomes are factories for protein synthesis. Many of these proteins form into complex cellular enzymes. The lysosomes are membranous sacs that store and release hydrolytic enzymes that digest macromolecules. The Golgi apparatus is an organelle consisting of membranous cisternae that form layered stacks. Endoplasmic vesicles, rich in glycoproteins, merge with the Golgi apparatus, increasing its membranous contents. Modified glycoprotein macromolecules from the Golgi are packaged and stored in the vesicles, which later pinch off from the trans-surface of the Golgi apparatus. The Golgi vesicles are transported to the plasma membrane, where they are involved in membrane repair and synthesis, or they may later modify to form lysosomes.

The endoplasmic reticulum (ER) is part of an extensive membranous transporting system within the cell associated with transport of proteins. On its outer surface the ER may be lined with ribosomes, which give the ER a granular appearance; or the surface of the ER may be free of ribosomes, giving the ER a smooth appearance. Ribosomes on the ER are factories for protein synthesis. Smooth ER secretes enzymes that facilitate the synthesis of phospholipids, hormones, carbohydrates, and hydrolytic enzymes, which break down drugs and toxic substances.

The semiautonomous mitochondria are double-membrane organelles with an outer, smooth membrane and an inner, extensively folded membrane that forms cristae. The cristae are lined with oxidative enzymes. Mitochondrial matrix, rich in oxidative enzymes, fills the lumen of the mitochondrion. Most of the ATP from glucose molecules is produced within the mitochondria by metabolic reactions.

Peroxisomes are membranous organelles rich in hydrolytic enzymes. Fatty acids, alcohol, hydrogen peroxide, and toxins are degraded to simpler molecules by peroxisomal enzymes.

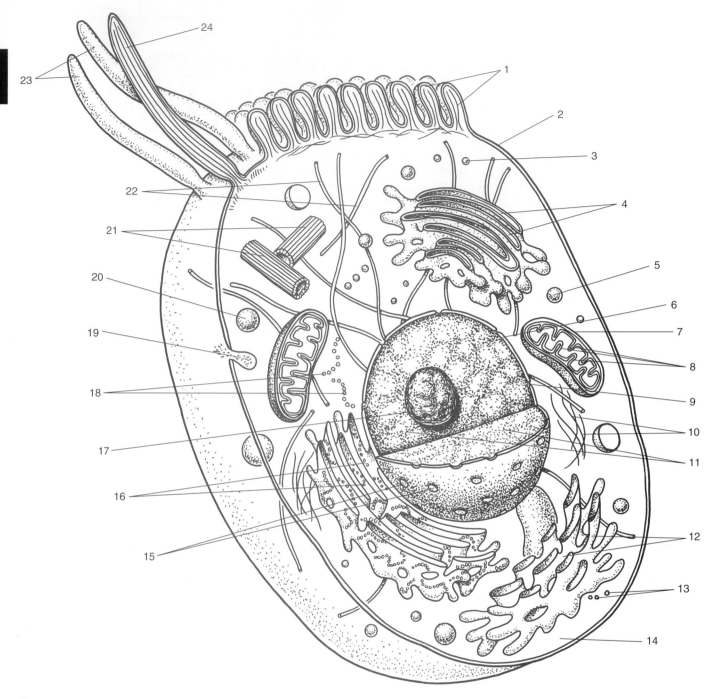

Figure 1.1   Diagrammatic presentation of an animal cell based on electron microscopic studies. Not all animal cells have cilia, microtubules, and microvilli. The structures shown in the illustration are drawn to facilitate the understanding of the animal cell.

1. Microvilli
2. Plasma (cell) membrane
3. Peroxisome
4. Golgi apparatus
5. Golgi secretory vesicle
6. Nuclear envelope
7. Outer mitochondrial membrane
8. Mitochondrial cristae
9. Nuclear pore
10. Microfilaments
11. Chromatin
12. Smooth endoplasmic reticulum (sER)
13. Inclusions
14. Cytoplasm
15. Nuclear pores
16. Rough endoplasmic reticulum (rER)
17. Nucleolus
18. Polysomes
19. Pinocytic vesicle
20. Lysosome
21. Centrioles
22. Microtubules
23. Cilia
24. Microtubules of cilia

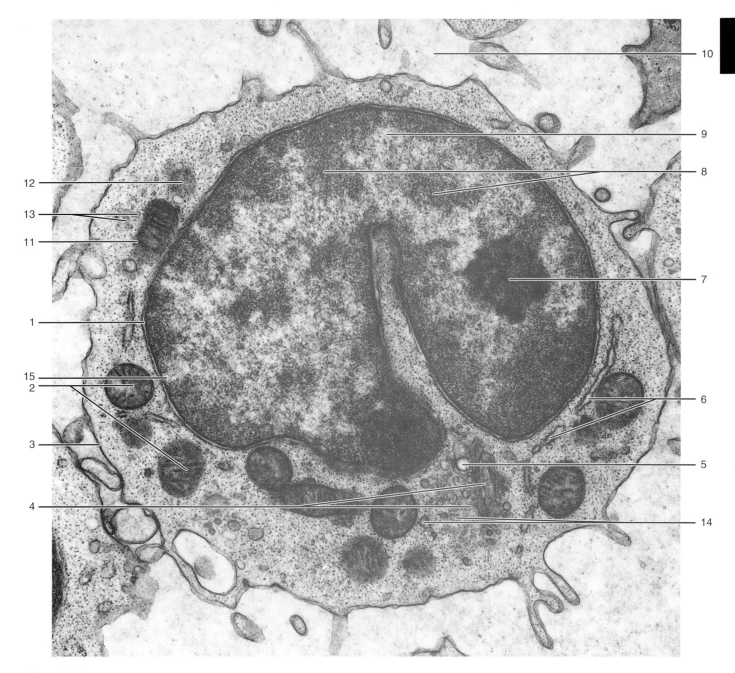

Figure 1.2   Transmission electron micrograph (TEM) of a white blood cell, demonstrating the cell structure and its organelles. (24,000×)

1. Nuclear envelope
2. Mitochondria
3. Plasma (cell) membrane
4. Golgi apparatus
5. Vacuole
6. Rough endoplasmic reticulum (rER)
7. Nucleolus
8. Heterochromatin
9. Euchromatin
10. Extracellular space
11. Mitochondrial cristae
12. Lysosome
13. Polysomes
14. Golgi vesicles
15. Nuclear pore

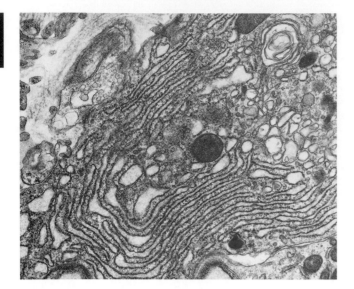

Figure 1.3 TEM of granular or rough endoplasmic reticulum (rER) from a kidney cell, demonstrating tubular lumen with dilated terminal extremities. Ribosomes line the cisternal membranes. Ribosomes are sites for protein synthesis. (36,000×)

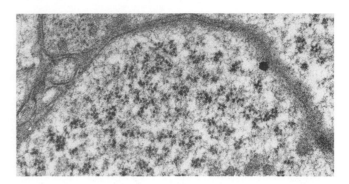

Figure 1.5 TEM of polyribosomes, which may exist as a rosette of ribosomes freely floating in the cytoplasm. Polysomes function as sites for protein synthesis. (60,000×)

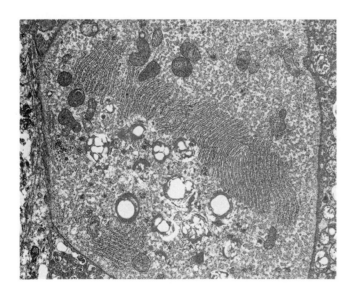

Figure 1.4 TEM of rough endoplasmic reticulum (rER), displaying ribosomes. (10,000×)

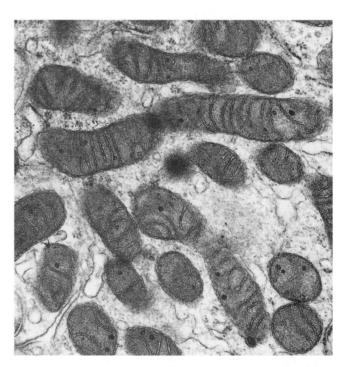

Figure 1.6 TEM of mitochondria, which are ovoid or rod-shaped in structure, with a smooth, outer investing membrane and an inner, extensively folded membrane. The inner-membrane folds, called cristae, are lined by oxidative enzymes. Mitochondria are involved in adenosine triphosphate (ATP) synthesis. (30,000×)

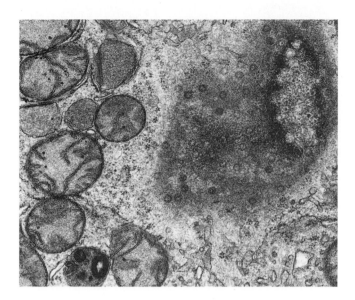

Figure 1.7 TEM of a cell displaying nucleus, nucleolus, nucleopores, and mitochondria at a higher magnification. The mitochondria show well-organized cristae and, in the matrix, opaque granules. (36,000×)

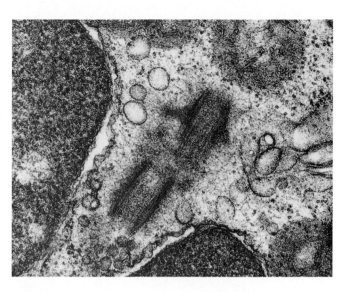

Figure 1.9 TEM of centrioles, which develop into hollow, blind cylinders approximately 160–560 nm long and 160–230 nm in diameter. Microtubules can be identified in the centrioles. (60,000×)

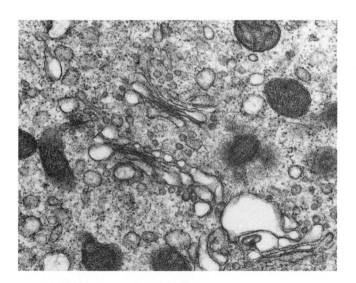

Figure 1.8 TEM of Golgi apparatus displaying curved, flattened, stacked saccules (cisternae). The distal end of the cisternae are dilated and form large sacs that later detach from the cisternae and float away as secretory vesicles. (60,000×)

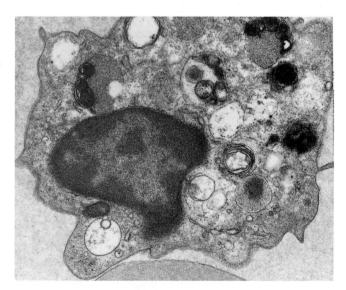

Figure 1.10 TEM of lysosomes, which are part of the enzymatic digestive system of the cell. Lysosomes vary in size, ranging from 50 nm in diameter for primary lysosomes, to 300 nm for secondary lysosomes. (18,000×)

Figure 1.11 TEM of a ciliated epithelial cell as seen in the nasal epithelium. Most of the cilia seen in the micrograph have been sectioned longitudinally. Also seen in the micrograph are large numbers of mitochondria. (1,500×)

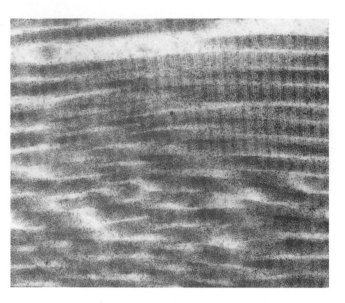

Figure 1.13 TEM of a cluster of collagen fibers in a longitudinal section. A collagen strand consists of fine microfibrils ranging from 45–100 nm in diameter. The microfibril bands repeat themselves at 64-nm intervals. (60,000×)

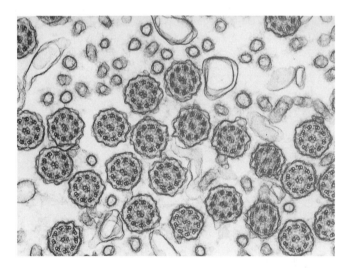

Figure 1.12 TEM of cilia in a cross section. A cilium shaft (axoneme) consists of nine fused, outer and two fused, central microtubules. This arrangement of microtubules is called the 9 + 2 pattern. (50,000×)

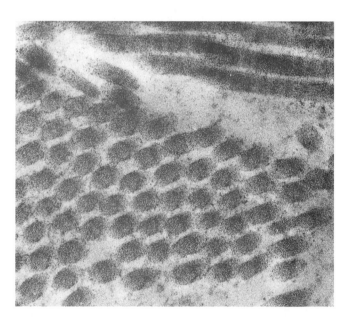

Figure 1.14 TEM of collagen fibers in a cross section. Each band is about 0.3–0.5 μm in diameter. Each fibril consists of microfibril subunits. (60,000×)

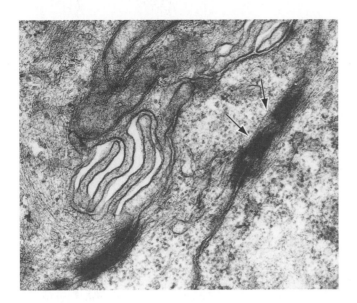

Figure 1.15   TEM of desmosomes, or macula adherens (see arrows). A desmosome consists of fine loops of tonofilaments that extend deep into the cytoplasm of adjoining cells. (15,000×)

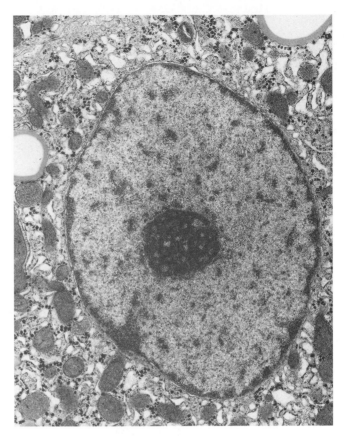

Figure 1.16   TEM of electron-dense glycogen granules as seen in a hepatocyte (liver cell). Glycogen is the principal form of carbohydrate reserve in liver, muscle, and adrenal cortex cells. The larger glycogen granules are generally arranged in clusters or in a rosette manner. (24,000×)

Figure 1.17   Light micrograph (LM) of human male chromosomal spread. The chromosomes have been isolated and photographed for karyotyping.

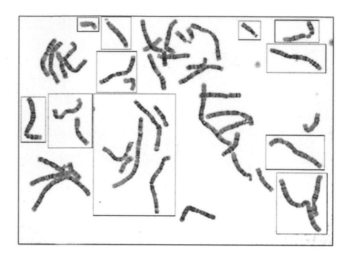

Figure 1.19   LM of human female mitotic metaphase chromosomes isolated and photographed for karyotyping.

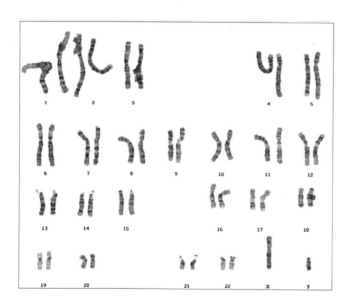

Figure 1.18   LM of human male chromosomes arrested at mitotic metaphase stage. The chromosomes are arranged to show 22 pairs of autosomes (see pairs 1–22), and a pair of XY sex chromosomes. The chromosomes are further classified into seven groups. Group A consists of chromosome pairs 1–3; Group B of pairs 4 and 5; Group C of pairs 6–12; Group D of pairs 13–15; Group E of pairs 16–18; Group F of pairs 19 and 20; and Group G of pairs 21 and 22.

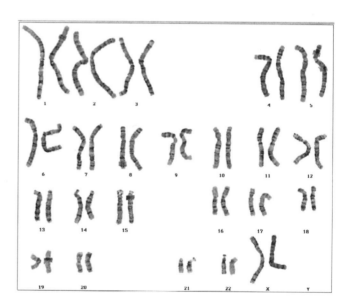

Figure 1.20   LM illustrating the human female chromosomes arrested at the metaphase stage of mitosis for karyotyping. Like the male, the female chromosomes are arranged in seven groups (see male karyotype, fig. 1.18). However, the female possesses 23 pairs of homologous chromosomes, with XX sex chromosomes.

The Cell Cycle

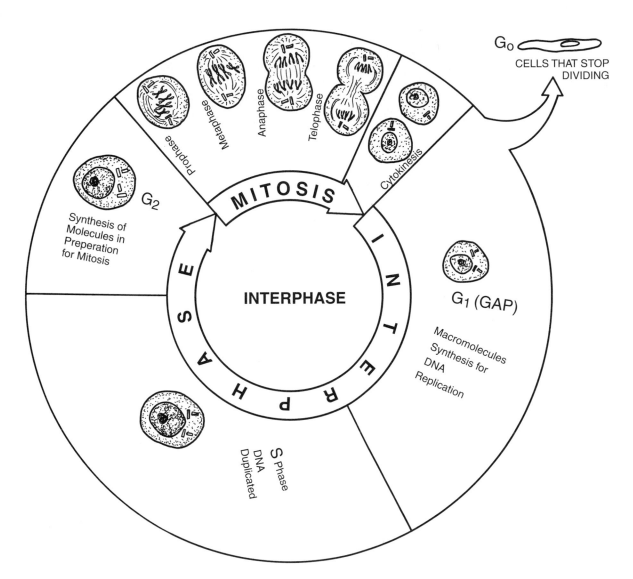

Figure 1.21   Diagrammatic presentation of the cell cycle, which includes the division of the nucleus, called mitosis, and division of the cytoplasm, called cytokinesis. These two divisions of the cell result in the formation of two daughter cells.

STAGES IN MITOSIS

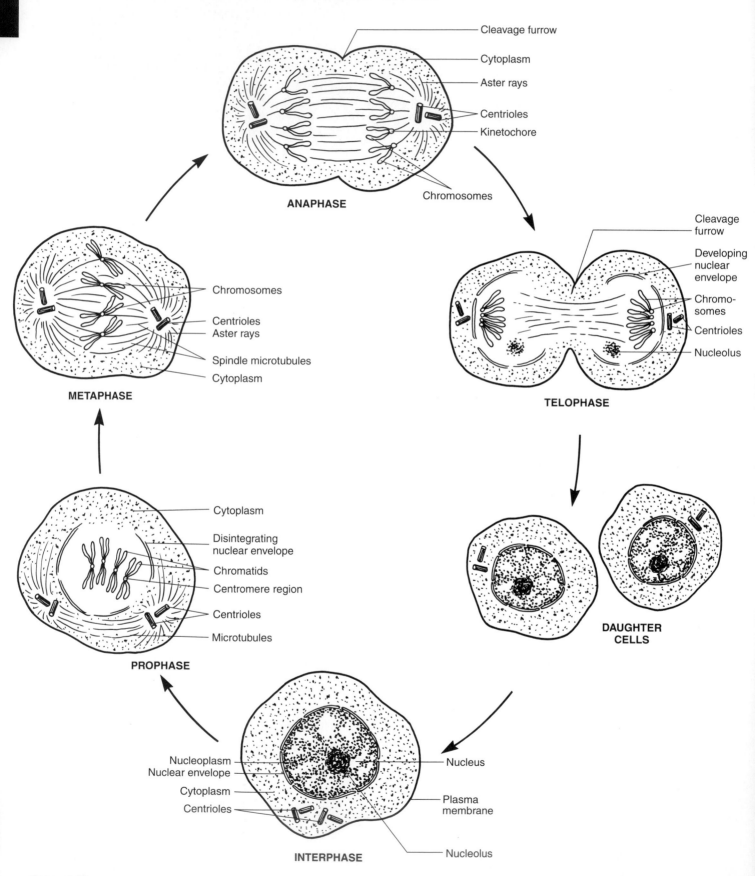

Figure 1.22   Diagrammatic presentation of stages in mitotic division as seen in animal cells.

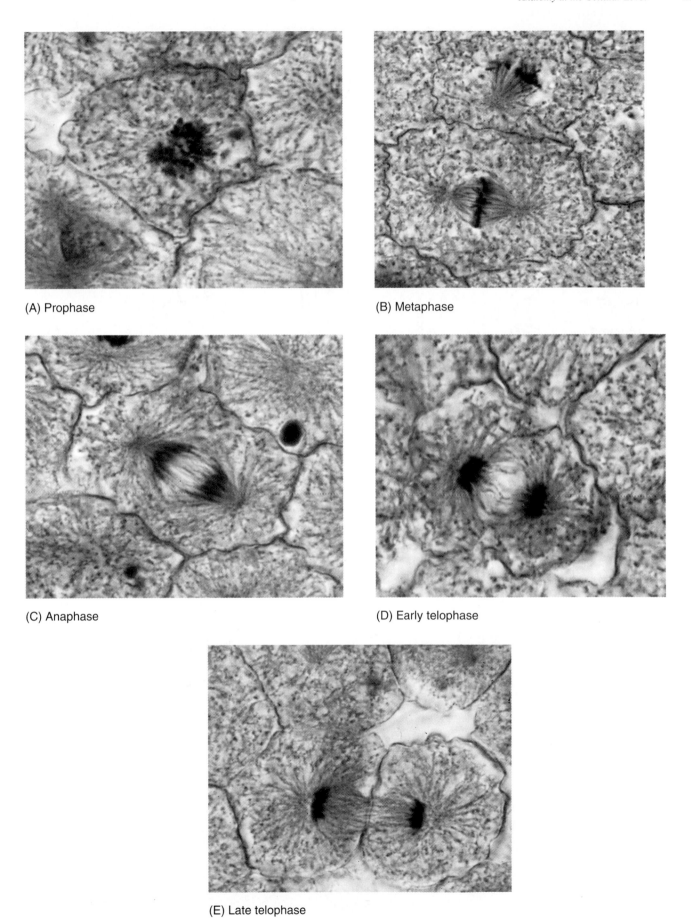

(A) Prophase

(B) Metaphase

(C) Anaphase

(D) Early telophase

(E) Late telophase

Figure 1.23   LM of stages in mitosis.

MEIOSIS—A DIAGRAMMATIC PRESENTATION

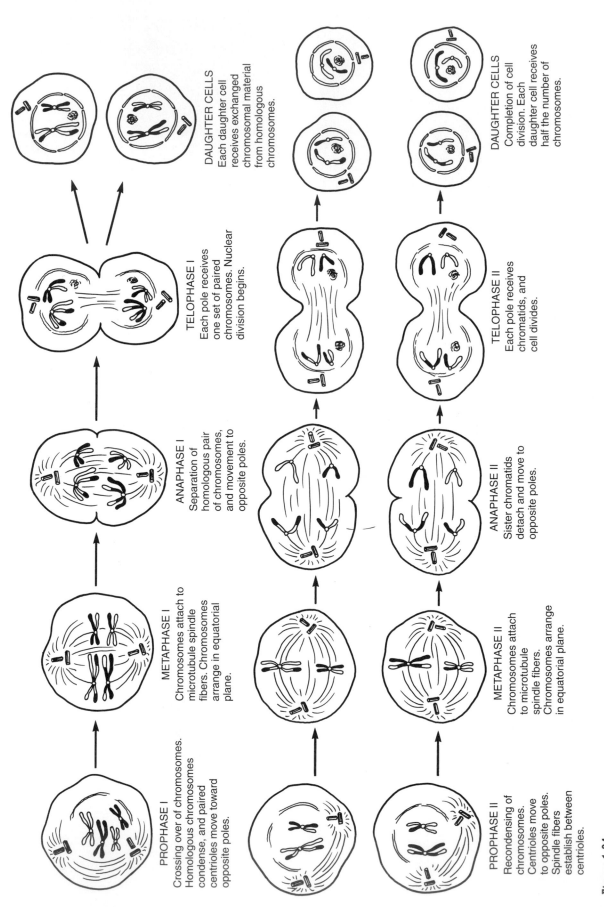

PROPHASE I
Crossing over of chromosomes. Homologous chromosomes condense, and paired centrioles move toward opposite poles.

METAPHASE I
Chromosomes attach to microtubule spindle fibers. Chromosomes arrange in equatorial plane.

ANAPHASE I
Separation of homologous pair of chromosomes, and movement to opposite poles.

TELOPHASE I
Each pole receives one set of paired chromosomes. Nuclear division begins.

DAUGHTER CELLS
Each daughter cell receives exchanged chromosomal material from homologous chromosomes.

PROPHASE II
Recondensing of chromosomes. Centrioles move to opposite poles. Spindle fibers establish between centrioles.

METAPHASE II
Chromosomes attach to microtubule spindle fibers. Chromosomes arrange in equatorial plane.

ANAPHASE II
Sister chromatids detach and move to opposite poles.

TELOPHASE II
Each pole receives chromatids, and cell divides.

DAUGHTER CELLS
Completion of cell division. Each daughter cell receives half the number of chromosomes.

Figure 1.24   Meiosis—A diagrammatic presentation

14

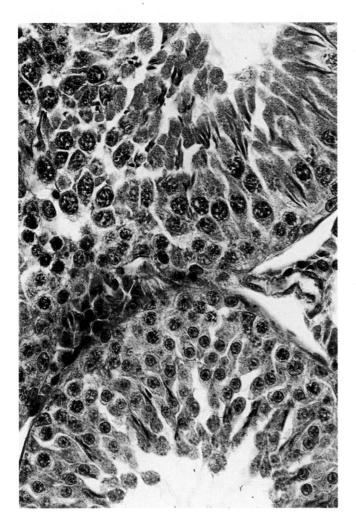

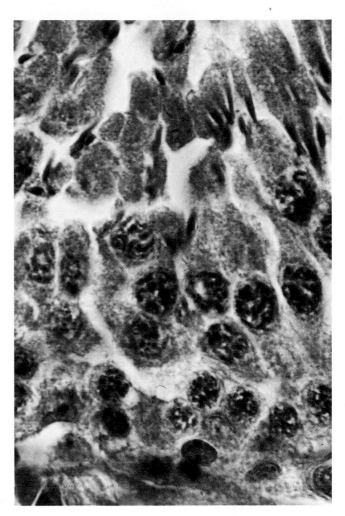

Figure 1.25 LM of stages in meiosis as seen in the seminiferous tubules of human testis. (400×)

Figure 1.26 LM of stages in meiosis at a higher magnification as seen in the seminiferous tubule of human testis. (1,000×)

CHAPTER TWO

Tissue Organization: Epithelial, Connective, Muscle, and Nervous Tissue

G roups of cells with similar morphology, functions, and embryonic origin form a given type of tissue. Groups of cells within tissues are separated by a thin veneer of connective tissue, or by some form of liquid material. Morphologically and functionally, the human body consists of four basic types of tissues: epithelial tissue, connective tissue, muscular tissue, and nervous tissue.

Epithelial tissue forms covering for the skin, organs of body systems, and lining for the body cavities. Invagination of the epithelium may lead to the formation of exocrine glands, such as sweat and sebaceous glands of the integument. Epithelial tissue also differentiates into some endocrine glands, involved in formation of some hormones.

Epithelial tissue rests upon the basal lamina, which separates the epithelium from the underlying connective tissue. Epithelial tissue, which borders membranes of the chest and abdominal cavity, blood vessels, and kidney corpuscles, may be a simple epithelium, with only a single layer of epithelial cells lining the basement lamina, or a stratified epithelium, with two to several layers of epithelial cells forming a covering or lining. Based on the cell type and cell arrangement, there are several other examples of epithelia, classified as simple squamous, simple columnar, simple cuboidal, stratified squamous, stratified cuboidal, stratified columnar (rare), transitional epithelium, and pseudostratified epithelium.

The connective tissue, a derivative of the embryological mesoderm, differentiates into connective tissue proper (e.g., areolar, regular, and adipose tissue), or a modified connective tissue such as cartilage, bone, and hemopoietic tissue. The intercellular substance that forms a homogeneous matrix of connective tissue is a combination of collagen fibers, chondromucoid, and associated glycosaminoglycan chemical complexes containing chondroitin 4-sulfate, chondroitin 6-sulfate, and keratin sulfate. Collagen fibers consisting of protein collagen are present in all forms of connective tissue. The thickness of the fibers varies from 1 to 12 microns (μm). Clusters of collagen fibers forming thick bundles, held together by mucoprotein, are not uncommon in connective tissue. Collagen fibers in the tissue are deposited by fibroblast, chondroblast, osteoblast, and odontoblast cells. The other two types of connective tissue fibers, the reticulate and elas-

tic, are found in specific connective tissue areas; for example, reticulate fibers are common around blood vessels, nerve fibers, muscle fibers, and adipose cells. Elastic fibers, composed of elastin, are generally found in loose fibrous connective tissue, in elastic cartilage, in ligaments, around blood vessels, and in embryonic tissue.

Connective tissue cells can be readily identified in areolar (loose) connective tissue, since the fibers are loosely arranged and the cells stand out against lightly staining matrix. Fibroblasts and macrophages make up most of the connective tissue cells. Other cells, such as mast cells, adipose cells, plasma cells, and occasional lymphocytes and eosinophils, can also be seen in connective tissue. Undifferentiated mesenchyme cells dominate the embryonic connective tissue.

Specialized connective tissue, such as cartilage and bone, exhibits many of the characteristics of connective tissue proper. Like connective tissue, cartilage and bone have connective tissue fibers at some stage of their development, similar types of cells, and ground substance that has been modified to give tensile strength. Chemically, the matrix in cartilage is concentrated with chondromucoid rich in chondroitin sulfates. In bone, matrix is reinforced by inorganic salts such as calcium and phosphates.

The muscular tissue is a specialized form of tissue designed to contract and relax and in the process produce movements. Based on its morphology and function, muscle tissue can be divided into three types: smooth and cardiac muscle, which is involuntary, and skeletal muscle, which is voluntary. Smooth muscle fibers are associated with the gastrointestinal tract, body walls of blood vessels, ducts of glands, tubular structures of the respiratory system, tubular structures of the urogenital system, larger lymphatic vessels, the dermis layer of the skin, and the iris and ciliary bodies of the eye. Smooth muscle morphology is not as discernible as that of skeletal and cardiac muscle, since smooth muscles lack organized myosin and actin bands.

The skeletal muscle fibers are covered by a thin sarcolemma. Within the cell, the sarcoplasm displays a large number of mitochondria, a well-established Golgi apparatus, granular sarcoplasmic reticulum, ribosomes, lysosomes, glycogen granules, and lipid vacuoles whose number increases as the muscle ages.

Under electron microscope, myofibrils in the skeletal muscle display smaller subunits of myofilaments: the isotropic actin and anisotropic myosin. The thick myosin filament, the A band, consists of a central H band. An M line bisects the H band. The thin actin I band is bisected by a narrow Z line. The muscle functional subunit extends from one Z line to the adjacent Z line and is termed a sarcomere. Seen in the sarcoplasm of a sarcomere are numerous mitochondria with closely packed cristae, a Golgi apparatus, ribosomes, granular endoplasmic reticulum, a few lysosomes, glycogen granules, and lipid vacuoles.

Morphologically, the nervous system can be divided into the peripheral and central nervous systems. The nervous tissue has a wide distribution in all organs of the body and includes the neurons, which form the basic functional units of the nervous system, and the neuroglia or supportive cells. The neuroglia cells are a conglomerate of cells such as the astrocytes, which transport nutrient molecules between neural cells and blood capillaries; ependymal cells, which form an epithelial lining for the ventricles of the brain and the neural canal of the spinal cord; oligodendrocytes and Schwann cells, which are associated with myelination of axon fibers; and microglia cells, which function as phagocytic cells for the nervous system.

Epithelial Tissue

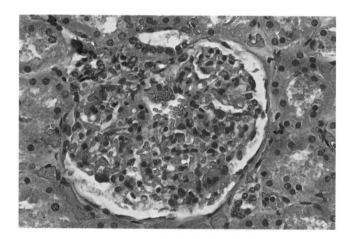

Figure 2.1   Light micrograph (LM) of kidney Bowman's capsule lined by single layer of squamous epithelial cells. In between the visceral and the parietal epithelial layers of the capsule lies the capsular space. (400×)

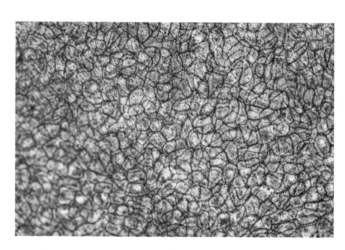

Figure 2.3   LM of simple squamous epithelial cells as seen in the surface view of mesothelium. (200×)

Figure 2.2   LM of lymph vessel lined by the endothelium, which consists of simple squamous epithelial cells. (400×)

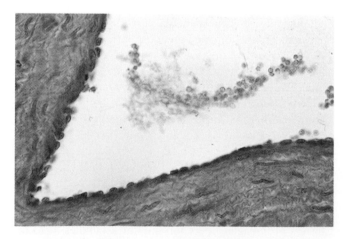

Figure 2.4   LM of simple squamous epithelial cells lining the umbilical cord. (400×)

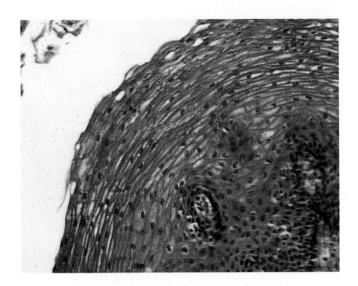

Figure 2.5   LM of inner lining of the esophagus displaying stratified squamous epithelium. (200×)

Figure 2.7   LM of stratified cuboidal epithelial cells present in the larger excretory ducts of submaxillary glands. Stratification of cells is generally confined to two or three layers (see arrow). (400×)

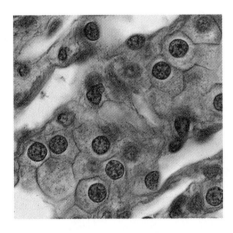

Figure 2.6   LM of simple cuboidal cells as seen in the tubules of kidney medulla. Note the large, centrally placed nucleus and the cube-shaped outer border of individual cells. (1,000×)

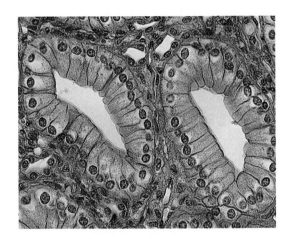

Figure 2.8   LM of simple columnar epithelium as viewed in the lining of papillary collecting tubules (ducts of Bellini) of the kidney. (400×)

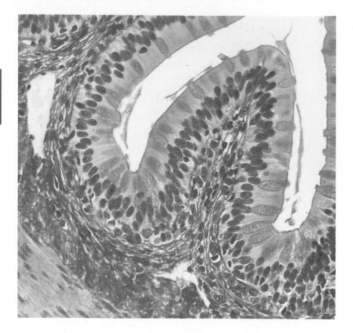

Figure 2.9   LM of simple columnar cells as seen in the lining of the small intestine. The columnar cells are tall with oblong nuclei and striated borders. (100×)

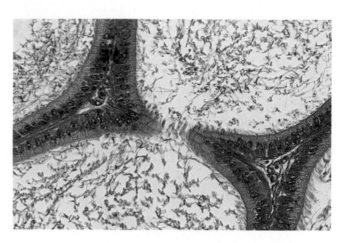

Figure 2.11   LM of stereocilia associated with columnar cells. Stereocilia are not mobile cilia. They are found only in the hair cells of the inner ear and in the epididymis of the testis. (400×)

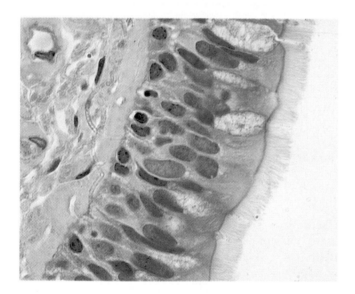

Figure 2.10   LM of pseudostratified ciliated columnar epithelium as observed in the inner lining of the trachea. The columnar cells appear to be stratified; however, all cells of the epithelium are in contact with the basement lamina. Also seen in the epithelium are goblet columnar cells. (1,000×)

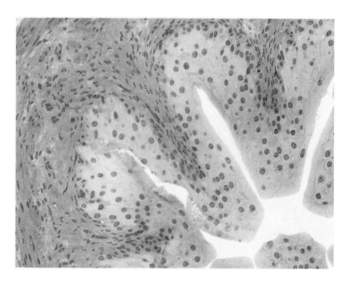

Figure 2.12   LM of transitional epithelium lining the urinary bladder. The cells are stratified with basal cells that are columnar or cuboidal in shape, whereas the cells of the middle layer are polyhedral in shape. The most superficial cells of the epithelium are large and dome-shaped and bulge into the lumen. (200×)

Connective Tissue

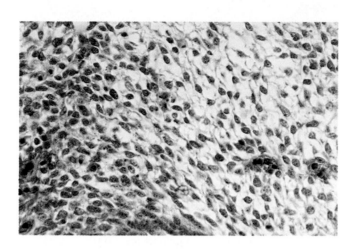

Figure 2.13   LM of undifferentiated embryonic mesenchyme connective tissue from a developing embryo. The cells are polymorphic with an oval nucleus. The gel-like matrix may include sparse reticular fibers. (400×)

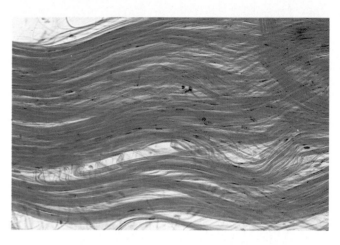

Figure 2.15   LM of collagen fibers as seen in fibrous connective tissue. At the molecular level, the fibers are composed of three alpha (α) chains of tropocollagen subunits wrapped around one another to form a helical configuration. (200×)

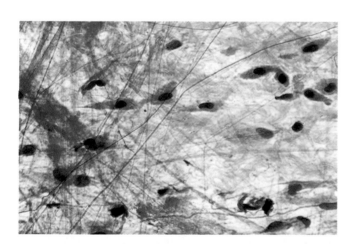

Figure 2.14   LM of areolar (loose) connective tissue. The tissue consists of cells, extracellular ground substance and fibers. (400×)

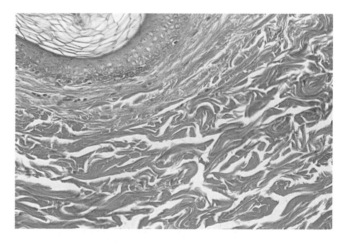

Figure 2.16   LM of dense, irregular connective tissue as seen in the dermis of the skin. (400×)

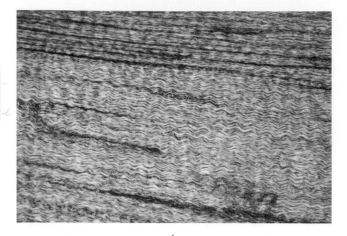

Figure 2.17 LM of dense, regular connective tissue fibers as seen in a tendon. The collagen fibers are densely packed and lie parallel to each other. Spindle-shaped fibroblast cells are located in between fibers. (200×)

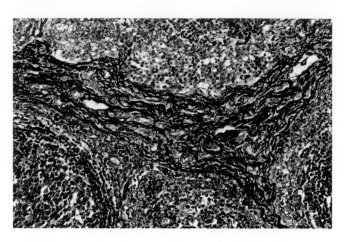

Figure 2.19 LM of reticulate connective tissue displaying a network of dense reticulate fibers. In between fibers lie concentrations of lymphocytes. (400×)

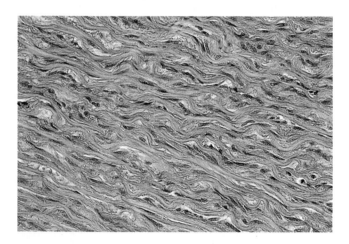

Figure 2.18 LM of dense, regular connective tissue at a higher magnification, demonstrating the arrangement of collagen fibers and fibroblast cells. (400×)

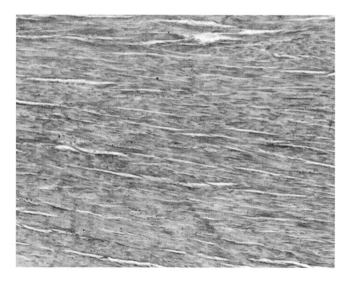

Figure 2.20 LM of dense, regular connective tissue as seen in ligamentum nuchae. (100×)

Cartilage

Bone

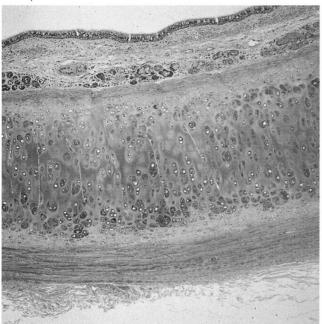

Figure 2.21 LM of hyaline cartilage. Note large ovoid chondrocytes in lacunae. At the upper and lower border of the cartilage lies the perichondrium and its chondrogenic cell layer. (100×)

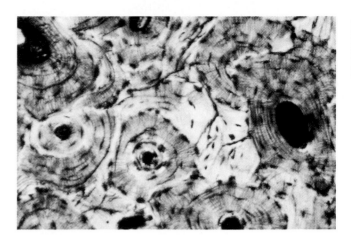

Figure 2.23 LM of ground bone. Seen in the micrograph are several Haversian systems. Each Haversian system has its own Haversian canal. (100×)

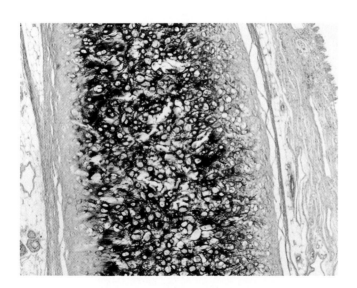

Figure 2.22 LM of elastic cartilage. The cartilage consists of dense bands of elastic fibers, chondrocytes in lacunae, and a bordering fibrous perichondrium. (400×)

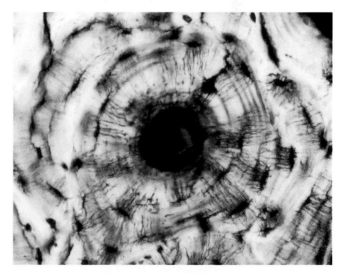

Figure 2.24 LM of a Haversian system at a higher magnification. Seen in the micrograph are concentric lamellae with lacunae, canaliculi, and a Haversian canal. (200×)

Blood

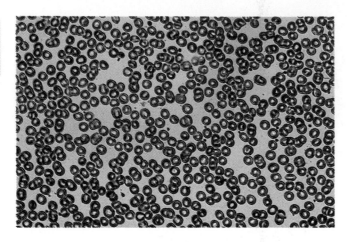

Figure 2.25   LM of erythrocytes as seen in peripheral blood. (400×)

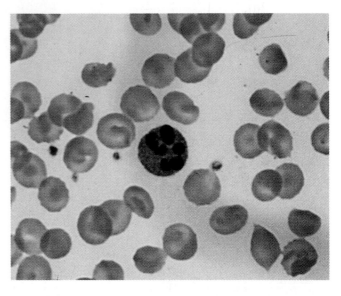

Figure 2.27   LM of a neutrophil. Note multilobed nucleus (polymorphonuclear leukocyte) and fine granules. (1,000×)

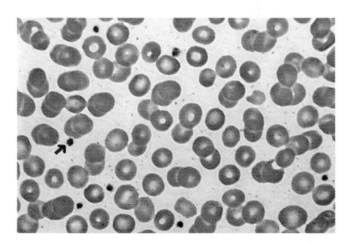

Figure 2.26   LM of erythrocytes and platelets (see arrow) at a higher magnification. (1,000×)

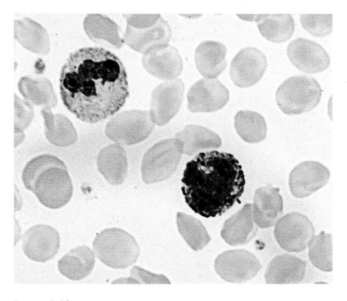

Figure 2.28   LM of a basophil. Note the S-shaped nucleus surrounded by large granules. (1,000×)

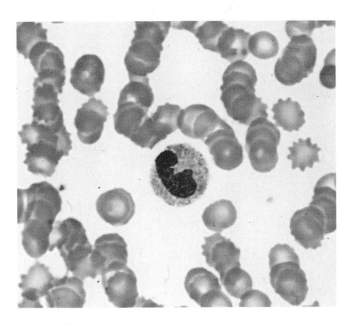

Figure 2.29   LM of an eosinophil. Note the bilobed nucleus surrounded by azurophilic granules. (1,000×)

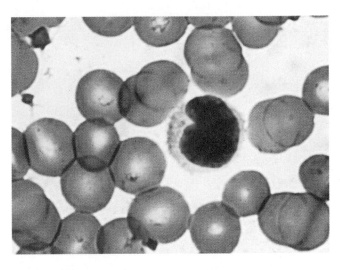

Figure 2.31   LM of a monocyte. Note the large kidney-shaped acentric nucleus, which occupies two-thirds of the cell. (1,000×)

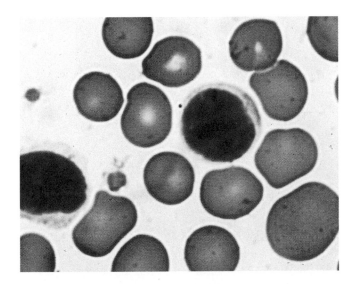

Figure 2.30   LM of a lymphocyte. Note the slightly indented, round, and acentrically placed nucleus, which occupies most of the cell. (1,000×)

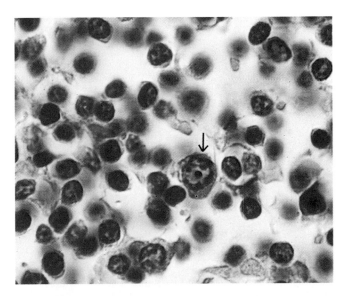

Figure 2.32   LM of a human bone marrow spread. Note the large, multinucleated megakaryocyte (see arrow) and blood cells at different stages of development. (1,000×)

Muscle

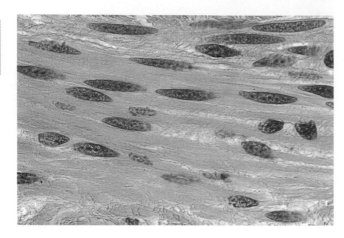

Figure 2.33   LM of smooth muscle, which lacks striations. The muscle fibers are fusiform, elongated, and taper at ends. Centrally located in the cell is an oval-shaped nucleus. (1,000×)

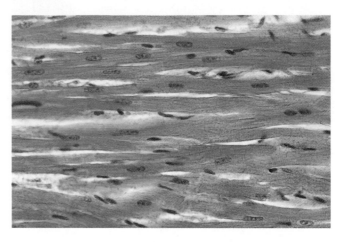

Figure 2.35   LM of cardiac muscle. The muscle fibers are branched, striated, and have a large, oval, centrally placed nucleus. The cells are separated by intercalated discs, which can be seen under high power. (400×)

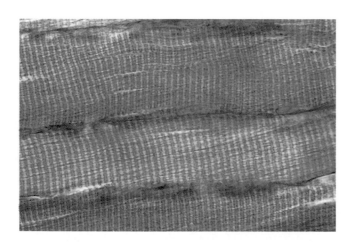

Figure 2.34   LM of skeletal muscle. The muscle fibers are long, multinucleated, and lie parallel to each other. Myofibrils in the muscle are composed of actin and myosin myofilaments, which can be identified as thin and thick bands, respectively. (1,000×)

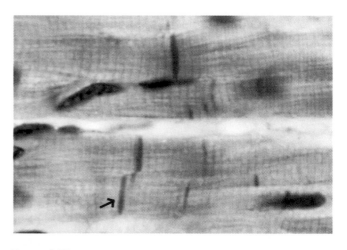

Figure 2.36   LM of cardiac muscle cells at a higher magnification, displaying striations and intercalated discs (see arrow). (1,000×)

Nervous Tissue

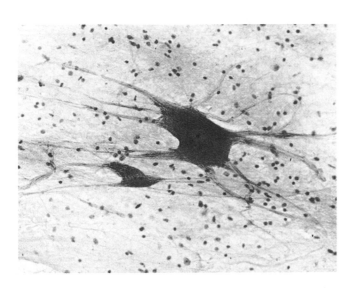

Figure 2.37   LM of multipolar neurons as seen in the gray matter of the spinal cord. The neurons are surrounded by neuroglia cells. (400×)

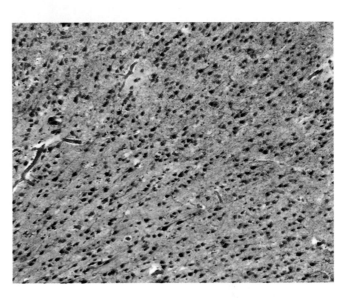

Figure 2.39   LM of neuroglia cells as seen in the cerebellar cortex. Surrounding the neuroglia cells are fibers from neurons in the cortical region. (100×)

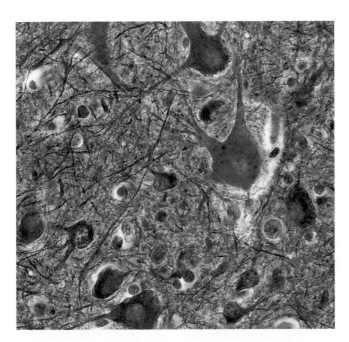

Figure 2.38   LM of pyramidal neurons found in the internal granular layer of the cerebral cortex. (1,000×)

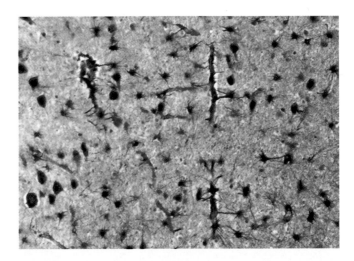

Figure 2.40   LM of fibrous astrocytes feeding off blood capillaries by perivascular feet. Also seen in the micrograph are myelinated axon fibers. (1,000×)

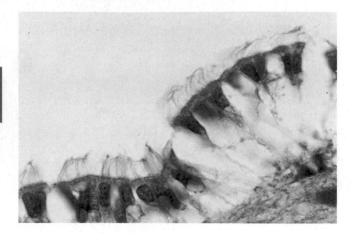

Figure 2.41   LM of cube-shaped ependymal cells, which line the ventricles of the brain and the central canal of the spinal cord. Extending from the cells are cilia that may be associated with movement of the cerebrospinal fluid. (1,000×)

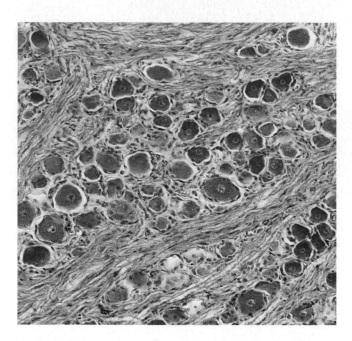

Figure 2.43   LM of the sensory ganglia at a higher magnification. The unipolar neurons display round cell bodies (perikaryon) with a centrally located nucleus. Surrounding the neurons are satellite cells and myelinated fibers. (200×)

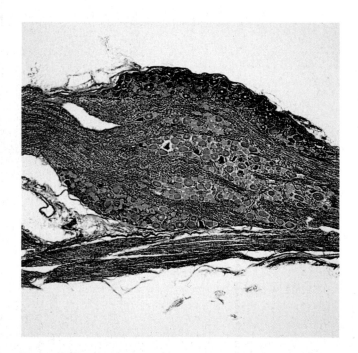

Figure 2.42   LM of a section through the sensory ganglia of the peripheral nervous system (PNS). Note the neuron fibers and surrounding satellite cells in the ganglia.

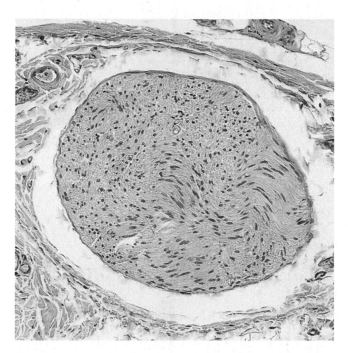

Figure 2.44   LM of a cross section through a peripheral nerve. Within the nerve, myelinated axons are surrounded by connective tissue called endoneurium. (200×)

Organ Systems

CHAPTER THREE

Integumentary System

The integumentary system includes the skin, which covers the body, and specialized derivatives of skin such as the glands, nails, and hair. The skin forms a protective barrier against microorganisms, regulates body temperature, excretes metabolic waste products, and acts as one of the sense organs for the body. Morphologically, the skin can be divided into two layers: the ectodermal-derived epidermis, and the mesodermal-derived vascular dermis and hypodermis. In between the epidermis and dermis lies the noncellular basement lamina.

The epidermis (epithelium) forms the upper layer of the skin and consists of several layers of stratified squamous cells. Several specialized cells can be identified in the epidermis. Five morphologically different layers of cells make up the epithelium. These layers include the lowermost stratum germinativum, followed by the stratum spinosum, the stratum granulosum, and the stratum corneum. In thick skin, such as the skin of the palm and foot, an added layer, the stratum lucidum, can be observed between the stratum granulosum and the stratum corneum.

Below the epidermis lies the dermis. The dermis is a highly vascular layer, and has a gelatinous matrix in which there is an extensive network of irregular collagen and elastic connective tissue fibers. In between fibers, fibroblasts, macrophages, mast cells, plasma cells, blood and lymph vessels, nerve and nerve endings, specialized receptors, and sebaceous and sweat glands are fairly common.

The hypodermis, also called superficial fascia, lies inferior to the dermis. The fascia is composed of adipose tissue surrounded by areolar connective tissue. The hypodermis of the eyelids, scrotum, and penis is free of adipose tissue.

Modified skin appendages, such as the nails, hair, and sweat and sebaceous glands, are also derived from the ectoderm. The nails are highly keratinized plates that lie over the dorsal surface of digits toward their distal end. The nail consists of the nail root, the nail bed, the nail plate, the nail fold, the eponychium, and the hyponychium. The hair follicles are complex modified skin appendages. The follicles consist of a connective tissue sheath, which itself consists of an inner glossy membrane, a middle connective tissue layer, an outer layer of mixed elastic and collagen fibers, covered by an outer and inner root sheath that blends with the epidermis.

The hair shaft, which may extend beyond the skin as hair, is composed of an inner medulla surrounded by a cortex and an outer layer of keratinized cuticle cells.

Also present in the integument are two types of exocrine glands: the sweat glands (eccrine and apocrine glands), and sebaceous glands that lie deep in the dermis layer. Secretions from these exocrine glands are transported by secretory ducts that open on the surface of the skin.

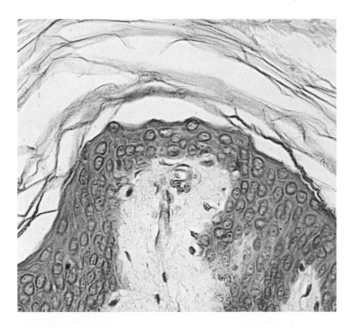

Figure 3.1   Light micrograph (LM) of human nonpigmented skin. Seen in the micrograph is the epidermis and the underlying dermis. Disintegrating stratum corneum has lost its cellular boundaries. (400×)

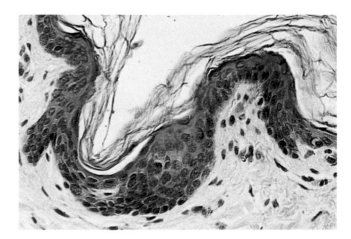

Figure 3.2   LM of pigmented skin. Seen in the micrograph are pigmented cells of stratum granulosum, and the upper layers of stratum spinosum. A few overlying stratum corneum cells can also be identified. The dermis layer lies below the epidermis. (400×)

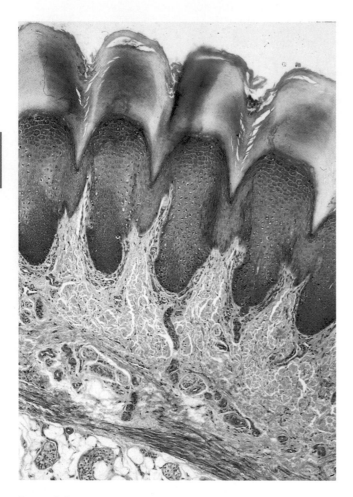

Figure 3.3   LM of thick keratinized skin. The epidermis is thick, with the lowermost single layer of cells forming the stratum germinativum, while the overlying cells form the stratum spinosum, spinosum granulosum, stratum lucidum, and the outermost stratum corneum. Below the epidermis lies the dermis. (100×)

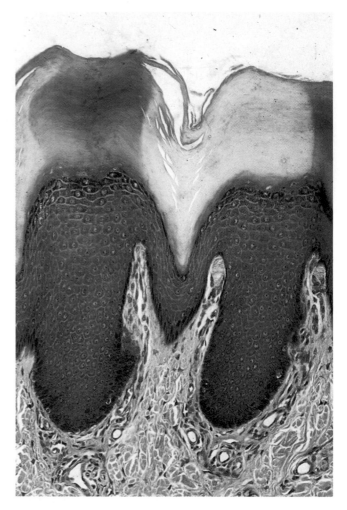

Figure 3.4   LM of thick keratinized skin at a higher magnification, showing the dermal papillae and the enclosed Meissner's corpuscles within the papillae. A few sweat glands can also be seen in the dermis. (200×)

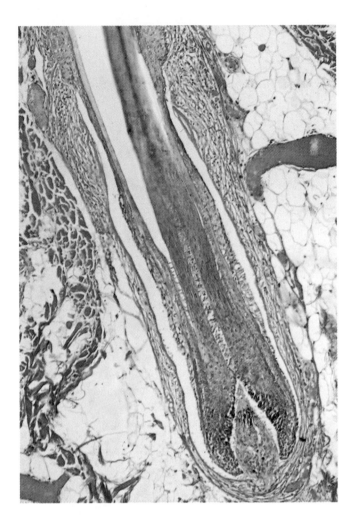

Figure 3.5 LM of a longitudinal section of hair shaft, hair follicle, and surrounding connective tissue of the dermis. The hair root and papilla of the follicle can also be identified in the micrograph. (100×)

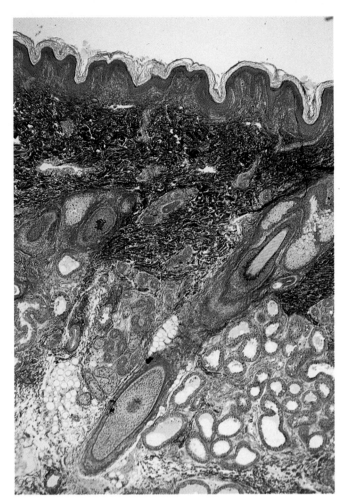

Figure 3.7 LM of skin in the axilla region. Seen in the micrograph is a developing hair shaft and a developing hair follicle. Sebaceous glands, apocrine sweat glands, and dense collagen fibers are prominent in the dermis. (40×)

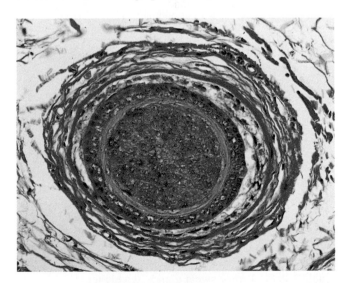

Figure 3.6 LM of a cross section through a hair follicle. The innermost medulla is surrounded by a cortex, inner Huxley's and outer Henle's layer, and the outermost layer of cuticle cells. A dermal sheath surrounds the follicle. (200×)

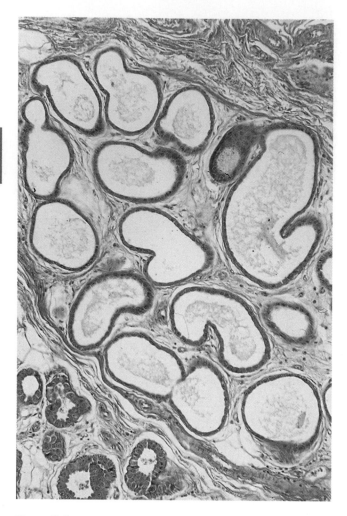

Figure 3.8   LM of a section through the skin (dermis) in an axilla region. Note apocrine sweat glands with large lumen. These glands are characteristic of skin in the axilla and genital regions. (100×)

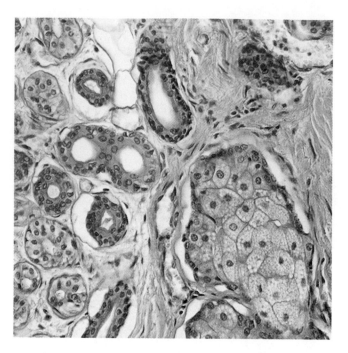

Figure 3.9   LM of a section through the skin (dermis), displaying sweat glands with large lumen and highly compact sebaceous glands. (200×)

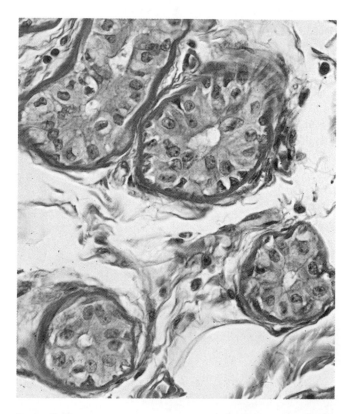

Figure 3.10   LM of sweat glands at a higher magnification. (200×)

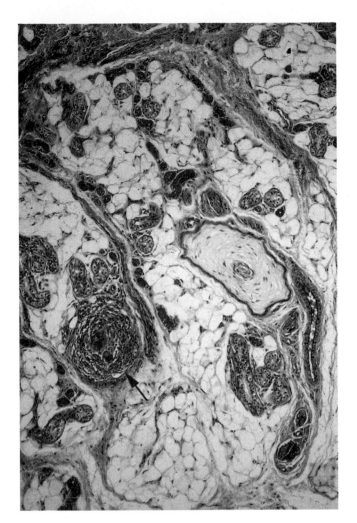

Figure 3.11   LM of a section through the skin. The dermis shows a glomus (an arteriovenous shunt) see arrow, and a large Pacinian corpuscle surrounded by adipose tissue of the hypodermis. (100×)

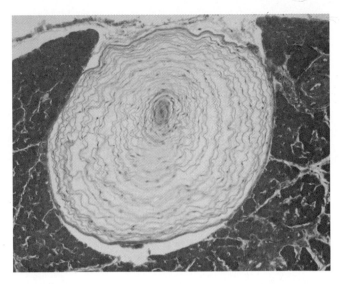

Figure 3.12   LM of a Pacinian corpuscle at a higher magnification, as seen in the pancreas. The function of the corpuscle in the pancreas is not clear. (200×)

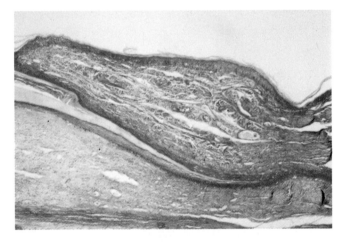

Figure 3.13   LM of a sagittal section through a fetal finger. On the anterior dorsal surface lies the specialized, keratinized appendage, the nail. Also seen in the micrograph is the nail fold, eponychium (cuticle), nail body, and free edge of the nail. (100×)

CHAPTER FOUR

Skeletal System

The skeletal system is composed of connective tissue, cartilage, and bone tissue. However, cartilage and bone are considered modified connective tissue since, at early stages of their development, a connective tissue framework is essential for establishing a dense, hard tissue that will later become a mature cartilage or bone. The cartilage and bone differ from the connective tissue by the chemical nature of their ground substance, which is designed for strength and rigidity. In cartilage tissue, the dense matrix is composed of chondromucoids with high concentrations of chondroitin sulfates; whereas bone matrix is mostly a combination of calcium phosphate, calcium carbonate, calcium fluoride, and magnesium fluoride.

In early stages of fetal development, cartilage and connective tissue lay down a temporary framework for the formation of bones (except flat bones of the body). As the fetus develops, most of the cartilage, which is hyaline cartilage, is replaced by bone except for a few places, such as in the respiratory system, articulating surfaces of bones, parts of the ear and nose, and the costal cartilage of the thorax.

In an adult mature skeleton three types of cartilage can be identified: hyaline cartilage, elastic cartilage, and the fibrocartilage. Hyaline cartilage is more common than the other two. The cartilage cells, chondrocytes, are confined to the cartilage matrix in small cavities called lacunae. The cartilage as a whole is surrounded by perichondrium, which forms a rigid layer of dense connective tissue infiltrated by chondroblast and fibroblast cells. The articular surfaces of bones lack a perichondrium. Instead, they are lined by a thin veneer of hyaline cartilage.

Bone, also called osseous tissue, is a tough, dense form of modified connective tissue that constitutes most of the skeletal system. In the skeletal system, there are two types of bones: the compact, or dense, bone, and cancellous, or spongy, bone. The bones differ merely by relative amount of dense matrix and empty spaces in their matrix. The bone cells—osteoblasts, osteocytes, and osteoclasts—can be found in all mature bone tissue. Osteocytes are osteoblast cells that were trapped in their matrix during the process of ossification. Osteocytes communicate with each other by extensions of their membrane and cytoplasm channels called canaliculi. The canaliculi facilitate the movement of metabolites from the blood capillaries to the osteocytes in lacunae.

Osteoclast cells are, functionally and morphologically, different from osteoblasts and osteocytes. Osteoclasts are multinucleated cells attached to the surface of the bone in shallow cavities known as Howship's lacunae. Functionally, the osteoclasts are involved in bone reabsorption that helps to maintain the necessary concentration of bone density in an adult skeleton. Parathormone hormone from the parathyroid gland stimulates the osteoclast cells to initiate the process of bone reabsorption.

Depending on the shape and function of a given bone, the bones are classified as long, short, irregular, or flat bones. The long bones include bones of the arm and legs; short bones are associated with wrists and ankles; irregular bones are odd-shaped bones, as seen in the vertebrae and some of the facial bones; and flat bones are found in the skull, scapulae, and the ribs.

Further, based on the arrangement of bones, the skeleton can be divided into an axial and an appendicular skeleton. The axial skeleton consists of the skull bones, hyoid bone, the vertebral column, which includes the cervical, thoracic, and lumbar vertebrae, the sacral and coccyx bones, the ear ossicles, and the bones of the thoracic cage, which include the ribs and sternum.

The appendicular skeleton consists of the pectoral girdles, clavicles, upper limb bones, the humerus, radius, ulna, carpals, metacarpals, phalanges, bones of the lower appendages, which include the pelvic girdle or coxal bones, the femur, tibia, fibula, patella, tarsal, metatarsal, and phalanges.

Skeletal Tissue

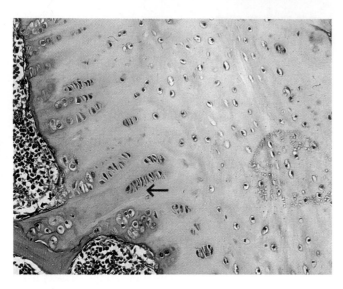

Figure 4.1   Light micrograph (LM) of fibrocartilage from pubic symphysis of pelvic girdle. Unlike hyaline and elastic cartilage, the fibrocartilage does not possess a perichondrium. Chondrocytes in fibrocartilage are arranged in rows (arrow) with alternating bundles of collagen fibers. (200×)

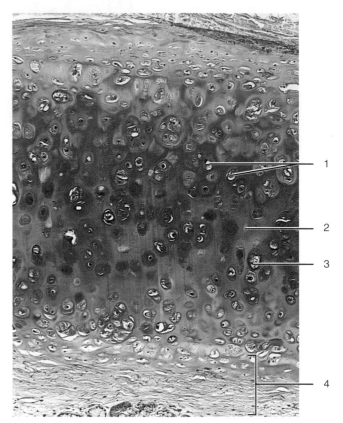

Figure 4.3   LM of hyaline cartilage in cross section. Note lacunae with chondrocytes throughout the homogenous matrix. Also seen in the micrograph is the perichondrium. (100×)

1. Lacunae   3. Chondrocyte in lacuna
2. Matrix of the cartilage   4. Perichondrium

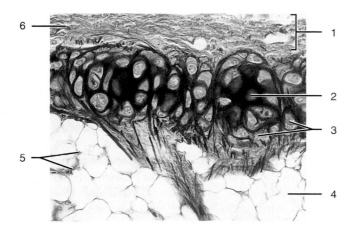

Figure 4.2   LM of elastic cartilage. Dark clusters of elastic fiber surrounding the lacunae, chondrocytes in the lacunae, and perichondrium on the upper surface of the cartilage are characteristics of elastic cartilage. (400×)

1. Perichondrium   4. Adipose tissue
2. Lacuna   5. Nuclei of adipocytes
3. Chondrocytes in lacunae 6. Fibroblast

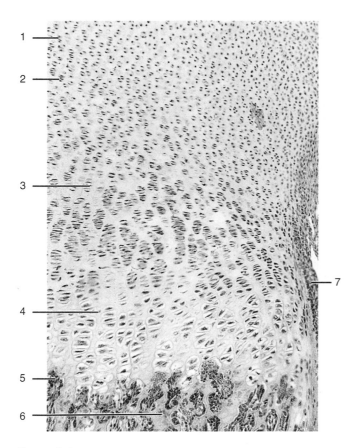

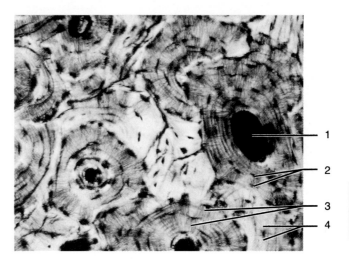

Figure 4.5 LM of ground bone as seen in a cross section. Haversian systems or osteons make up the dense structural units of bone. (100×)

1. Haversian canal
2. Lacunae with osteocytes
3. Canaliculi
4. Bone matrix

Figure 4.4 LM of zones of elaborating cartilage and developing bone at the junction of epiphysis and diaphysis of a developing long bone. (100×)

1. Zone of reserve cartilage
2. Zone of proliferating cartilage
3. Zone of hypertrophying cartilage cells
4. Zone of calcifying cartilage
5. Bone trabecula
6. Bone marrow
7. Periosteum

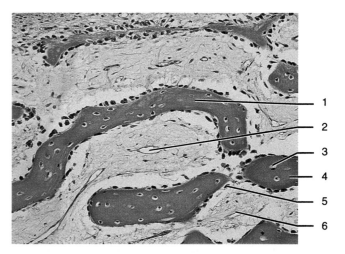

Figure 4.6 LM of bone trabeculae. Lining the trabeculae are bone-depositing osteoblast cells. (200×)

1. Trabecula
2. Blood vessel
3. Osteocyte in a lacuna
4. Matrix of developing bone
5. Osteoblast cell
6. Surrounding connective tissue

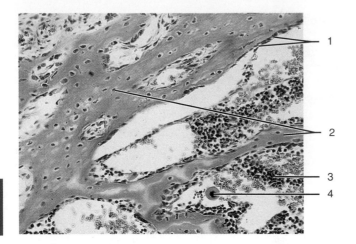

Figure 4.7   LM of cancellous or spongy bone. Note bone-marrow filled spaces between the trabeculae. (200×)

1. Osteoblasts
2. Trabeculae
3. Bone marrow
4. Developing leukocyte

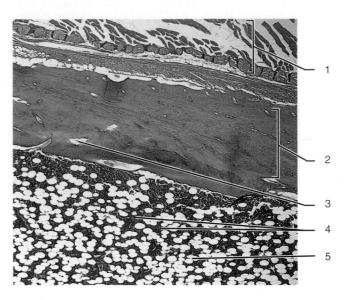

Figure 4.8   LM of decalcified bone. Note the periosteum in the upper part of the micrograph, and the bone marrow in the lower part. In between the bone marrow lies the adipose tissue. (100×)

1. Periosteum
2. Decalcified bone
3. Lacuna with osteocyte
4. Bone marrow
5. Adipose tissue

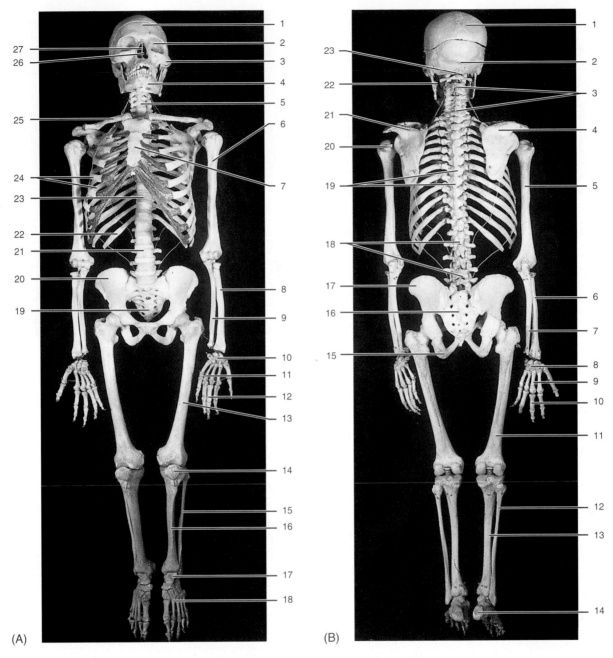

Figure 4.9 Adult skeleton. (a) anterior view, (b) posterior view.

1. Frontal bone
2. Orbit
3. Zygomatic bone
4. Mandible
5. Cervical vertebra
6. Humerus
7. Sternum
8. Radius
9. Ulna
10. Carpal bones
11. Metacarpal bones
12. Phalanges
13. Femur
14. Patella

15. Fibula
16. Tibia
17. Tarsal bones
18. Metatarsal bones
19. Sacrum
20. Ilium
21. Lumbar vertebra
22. Floating rib
23. Thoracic vertebra
24. Ribs
25. Clavicle
26. Nasal cavity
27. Nasal septum

1. Parietal bone
2. Occipital
3. Cervical vertebra
4. Scapula
5. Humerus
6. Radius
7. Ulna
8. Carpal bones
9. Metacarpal bones
10. Phalanges
11. Femur
12. Fibula

13. Tibia
14. Calcaneus
15. Ischium
16. Sacrum
17. Ilium
18. Lumbar vertebra
19. Thoracic vertebra
20. Head of humerus
21. Spine of scapula
22. Mandible
23. Atlas vertebra

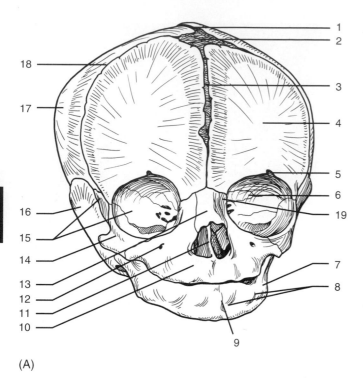

(A)

(B)

Figure 4.10   Fetal skull—anterior view.

1. Sagittal suture
2. Anterior fontanelle
3. Frontal suture
4. Frontal bone
5. Supraorbital foramen
6. Lesser wing of sphenoid bone
7. Mandible
8. Elevations of deciduous teeth
9. Symphysis of mandible
10. Maxilla

11. Nasal septum
12. Infraorbital foramen
13. Nasal bone
14. Zygomatic bone
15. Greater wing of sphenoid bone
16. Temporal bone
17. Parietal bone
18. Coronal suture
19. Frontal process of maxilla

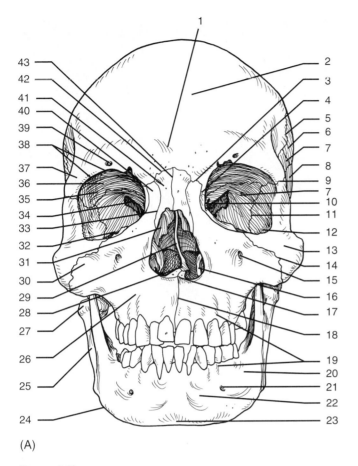

(A)

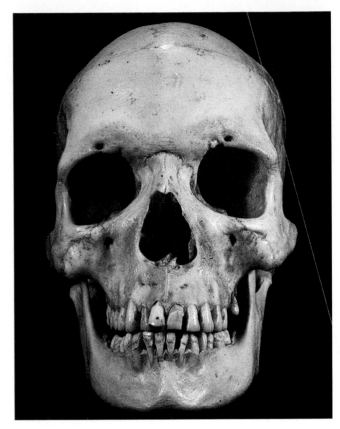

(B)

Figure 4.11 Human skull—anterior view.

1. Glabella
2. Frontal bone
3. Frontomaxillary suture
4. Frontolacrimal suture
5. Coronal suture
6. Parietal bone
7. Greater wing of sphenoid bone
8. Temporal bone
9. Frontozygomatic suture
10. Lacrimal bone
11. Sphenozygomatic suture
12. Infraorbital margin
13. Zygomatic bone
14. Zygomaticomaxillary suture
15. Infraorbital foramen
16. Inferior nasal concha
17. Anterior nasal spine
18. Intermaxillary suture
19. Alveolar process
20. Body of mandible
21. Mental foramen
22. Mental tubercle

23. Mental protuberance
24. Inferior border of mandible
25. Ramus of mandible
26. Maxilla
27. Mandibular condyle
28. Vomer
29. Perpendicular plate
30. Zygomatic process of maxilla
31. Middle nasal concha
32. Inferior orbital fissure
33. Nasomaxillary suture
34. Superior orbital fissure
35. Sphenofrontal suture
36. Orbital surface of frontal bone
37. Zygomatic process of frontal bone
38. Supraorbital margin
39. Supraorbital foramen
40. Frontal process of maxilla
41. Nasal bone
42. Frontonasal suture
43. Internasal suture

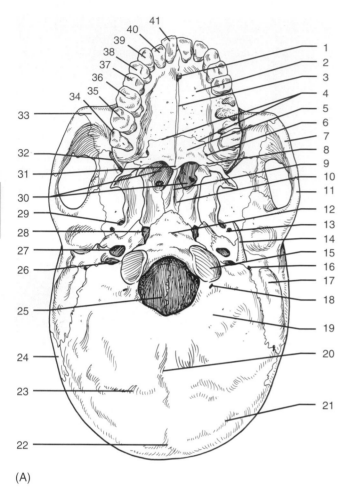

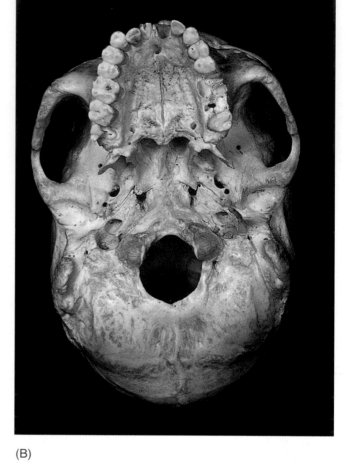

(A)   (B)

Figure 4.12   Base of human skull—external surface.

1. Incisive fossa
2. Maxillary palatine process
3. Median palatine suture
4. Transverse palatine suture
5. Horizontal plate of palatine bone
6. Greater palatine foramen
7. Zygomatic bone
8. Pyramidal process of palatine bone
9. Vomer
10. Ala
11. Zygomatic process of temporal bone
12. Pharyngeal tubercle
13. Foramen spinosum of sphenoid bone
14. Styloid process
15. Occipital condyle
16. Stylomastoid foramen
17. Mastoid process
18. Condyloid fossa
19. Occipital bone
20. External occipital crest
21. Superior nuchal line

22. External occipital protuberance
23. Inferior nuchal line
24. Parietal bone
25. Foramen magnum
26. Jugular foramen
27. Carotid canal
28. Foramen lacerum
29. Foramen ovale of sphenoid bone
30. Choanae (posterior nasal aperture)
31. Posterior nasal spine
32. Inferior orbital fissure
33. Zygomatic process of maxilla
34. Third molar
35. Second molar
36. First molar
37. Second premolar
38. First premolar
39. Canine
40. Lateral incisor
41. Central incisor

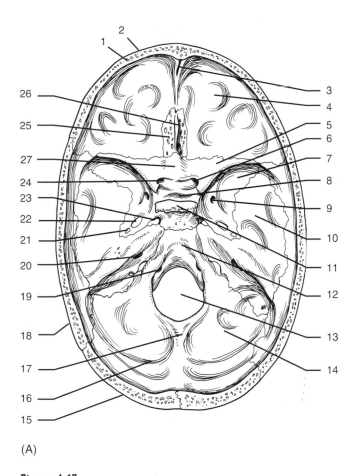

(A)

(B)

Figure 4.13   Cranial vault—superior view.

1. Diploë
2. Frontal bone
3. Frontal crest
4. Anterior cranial fossa
5. Lesser wing of sphenoid bone
6. Chiasmatic groove of sphenoid bone
7. Greater wing of sphenoid bone
8. Hypophyseal fossa of sella turcica
9. Foramen rotundum
10. Middle cranial fossa
11. Dorsum sellae of sella turcica
12. Clivus
13. Foramen magnum
14. Posterior cranial fossa
15. Occipital bone
16. Internal occipital protuberance
17. Internal occipital crest
18. Parietal bone
19. Hypoglossal canal
20. Jugular foramen
21. Foramen spinosum
22. Foramen ovale
23. Foramen lacerum
24. Carotid groove
25. Cribriform plate of ethmoid bone
26. Crista galli of ethmoid bone
27. Jugum of sphenoid bone

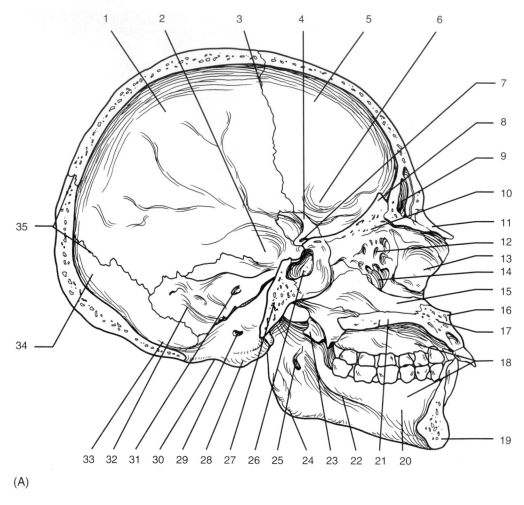

(A)

Figure 4.14 Left half of human skull—sagittal section.

1. Parietal bone
2. Middle cranial fossa
3. Coronal suture
4. Lesser wing of sphenoid
5. Frontal bone
6. Anterior cranial fossa
7. Optic canal of sphenoid
8. Crista galli of ethmoid
9. Frontal sinus
10. Cribriform plate of ethmoid
11. Nasal bone
12. Ethmoid bone
13. Perpendicular plate of ethmoid bone
14. Maxillary sinus
15. Vomer
16. Nasal spine of maxilla bone
17. Incisive canal of maxilla bone
18. Alveolar margin
19. Mental protuberance of mandible
20. Body of mandible
21. Palatine process of maxilla bone
22. Mylohyoid groove of mandible
23. Pterygoid process of sphenoid
24. Angle of mandible
25. Mandibular foramen
26. Sphenoidal sinus
27. Basilar part of occipital bone
28. Occipital condyle
29. Hypoglossal canal
30. Margin of foramen magnum
31. Internal acoustic meatus
32. Temporal bone
33. Posterior cranial fossa
34. Occipital bone
35. Lambdoid suture

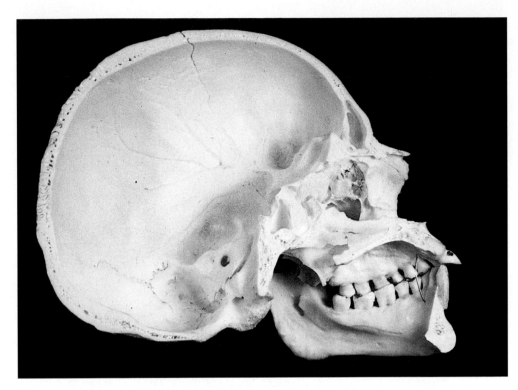

(B)

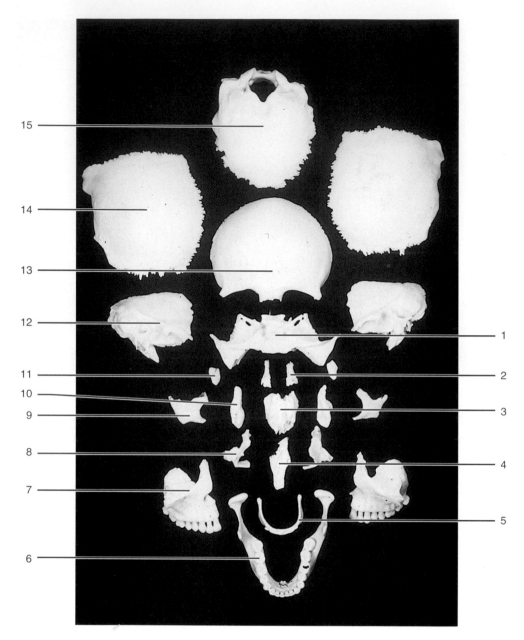

Figure 4.15   Disarticulated skull.

1. Sphenoid bone
2. Nasal bone
3. Ethmoid bone
4. Vomer
5. Hyoid bone
6. Mandible
7. Maxilla
8. Palatine bone
9. Zygomatic bone
10. Nasal concha
11. Lacrimal bone
12. Temporal bone
13. Frontal bone
14. Parietal bone
15. Occipital bone

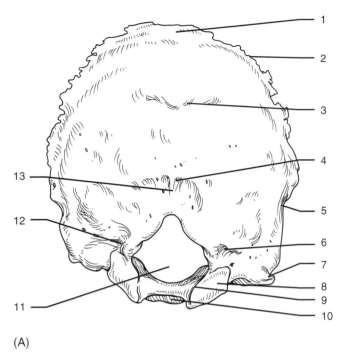

(A)

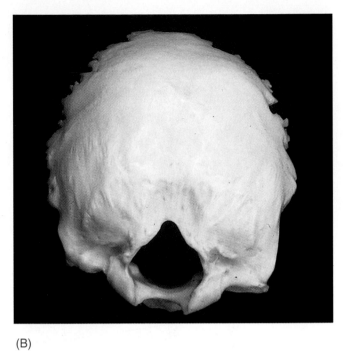

(B)

Figure 4.16   Occipital bone—posterior inferior view.

1. Squamous part of occipital bone
2. Lambdoid suture
3. External occipital protuberance
4. Inferior nuchal line
5. Occipitomastoid suture
6. Condylar fossa
7. Jugular process
8. Occipital condyle
9. Hypoglossal canal
10. Basilar part of squama
11. Foramen magnum
12. Condylar canal
13. External occipital crest

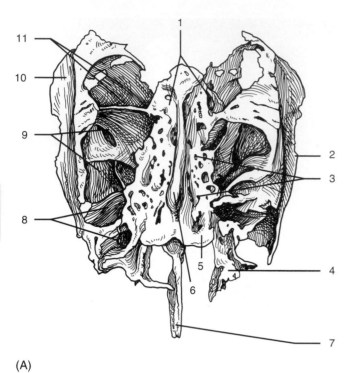

(A)

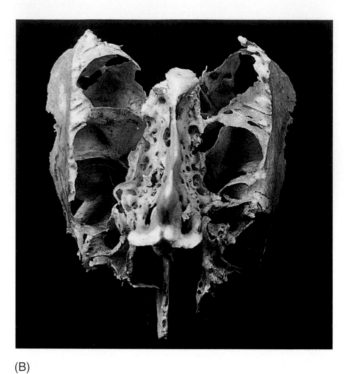

(B)

Figure 4.17   Ethmoid bone—superior view.

1. Cribriform (horizontal) plate
2. Ethmoidal labyrinth
3. Olfactory foramina
4. Middle nasal concha (turbinate)
5. Wings of crista galli (ala)
6. Crista galli
7. Perpendicular plate
8. Anterior ethmoidal air cells (sinus)
9. Middle ethmoidal air cells (sinus)
10. Orbital plate (lamina papyracea)
11. Posterior ethmoidal air cells (sinus)

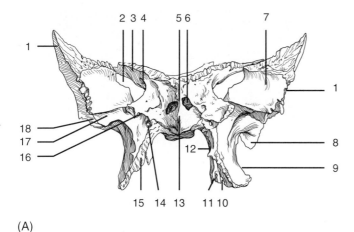

(A)

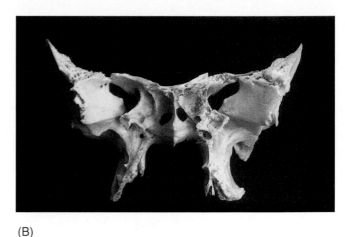

(B)

Figure 4.18   Sphenoid bone. (a) and (b), anterior view; (c) and (d), posterior view.

1. Temporal surface of greater wing
2. Superior orbital fissure
3. Lesser wing
4. Optic canal
5. Sphenoidal crest
6. Sphenoid sinus
7. Orbital surface of greater wing
8. Sphenoidal spine
9. Lateral plate of pterygoid process

10. Pterygoid notch
11. Pterygoid hamulus
12. Medial plate of pterygoid process
13. Sphenoid rostrum
14. Pterygoid canal
15. Pterygopalatine sulcus
16. Foramen rotundum
17. Pterygopalatine surface
18. Infratemporal surface

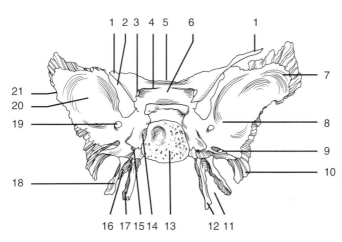

(C)

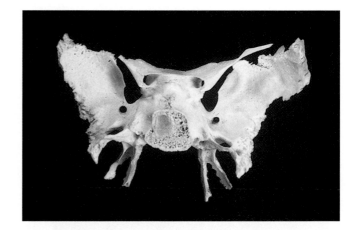

(D)

1. Lesser wing
2. Superior orbital fissure
3. Optic canal
4. Posterior clinoid process
5. Tuberculum sellae
6. Clivus
7. Cerebral surface of greater wing
8. Middle cranial fossa
9. Foramen ovale
10. Sphenoidal spine
11. Pterygoid notch

12. Pterygoid hamulus
13. Body of sphenoid bone (sectioned)
14. Carotid sulcus
15. Pterygoid canal
16. Pterygoid fossa
17. Medial pterygoid plate
18. Lateral pterygoid plate
19. Foramen rotundum
20. Greater wing
21. Squamous margin

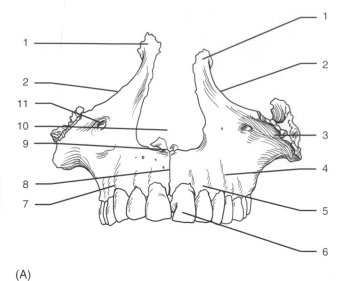

(A)

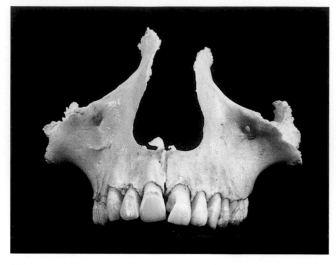

(B)

Figure 4.19   Maxilla with teeth—anterior view.

1. Frontal process
2. Infraorbital margin
3. Zygomatic process
4. Canine fossa
5. Alveolar process
6. Teeth

7. Juga alveolaria
8. Intermaxillary suture
9. Anterior nasal spine
10. Anterior nasal aperture
11. Infraorbital foramen

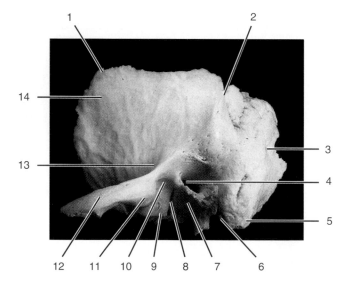

Figure 4.20   Temporal bone—lateral view.

1. Parietal margin
2. Parietal notch
3. Occipital margin
4. External acoustic meatus
5. Mastoid process
6. Styloid process
7. Tympanic part of temporal bone

8. Petrotympanic fissure
9. Apex of petrous part
10. Mandibular fossa
11. Articular tubercle
12. Zygomatic process
13. Groove for middle temporal artery
14. Squamous part of temporal bone

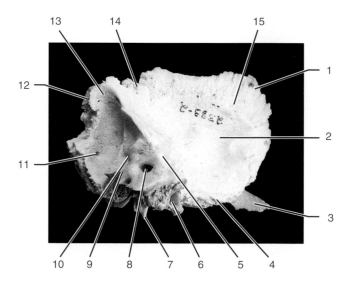

Figure 4.21   Temporal bone—medial view.

1. Parietal margin
2. Groove for meningeal vessel
3. Zygomatic process
4. Sphenoid margin
5. Trigeminal impression
6. Apex of petrous part
7. Styloid process
8. Internal acoustic meatus

9. Cochlear canaliculus
10. Aqueduct of the vestibule
11. Mastoid foramen
12. Occipital margin
13. Groove for sigmoid sinus
14. Parietal notch
15. Squamous part of temporal bone

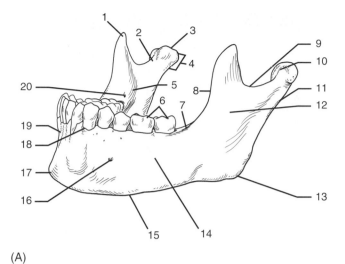

(A)

(B)

Figure 4.22 Mandible—lateral view.

1. Coronoid process
2. Neck
3. Head (caput)
4. Mandibular condyle
5. Mylohyoid groove
6. Teeth (molars)
7. Oblique line
8. Anterior border of ramus
9. Mandibular notch
10. Pterygoid fossa
11. Posterior border of ramus
12. Ramus of mandible
13. Mandibular angle
14. Body of mandible
15. Inferior border of mandible
16. Mental foramen
17. Mental protuberance
18. Alveolar process
19. Alveolar crest
20. Mandibular foramen

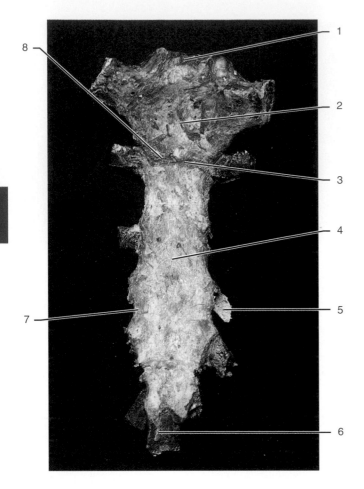

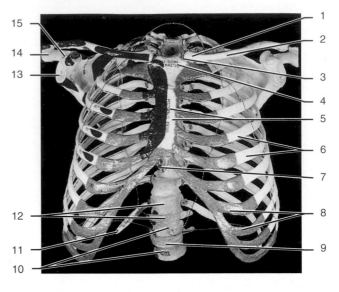

Figure 4.24   Shoulder girdle and thorax—anterior view.

1. Sternal end of clavicle   9. Vertebral disc
2. Clavicle   10. Lumbar vertebrae
3. Sternoclavicular joint   11. Floating rib
4. Manubrium of sternum   12. Thoracic vertebrae
5. Body of sternum   13. Glenoid cavity
6. Ribs   14. Acromion
7. Xiphoid process   15. Scapula
8. Costal cartilage

Figure 4.23   Sternum.

1. Jugular notch   5. Costal cartilage
2. Manubrium   6. Xiphoid process
3. Manubriosternal joint   7. Notch for costal cartilage
4. Body   8. Sternal angle

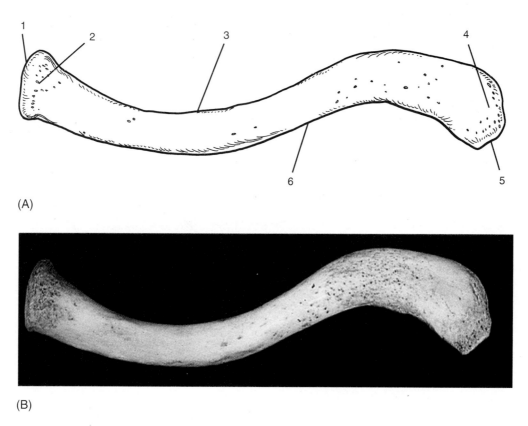

(A)

(B)

Figure 4.25   Clavicle—superior view, left side.

1. Articular facet for sternum
2. Sternal (medial end)
3. Dorsal border
4. Acromion (lateral end)
5. Articular facet for acromion
6. Ventral border

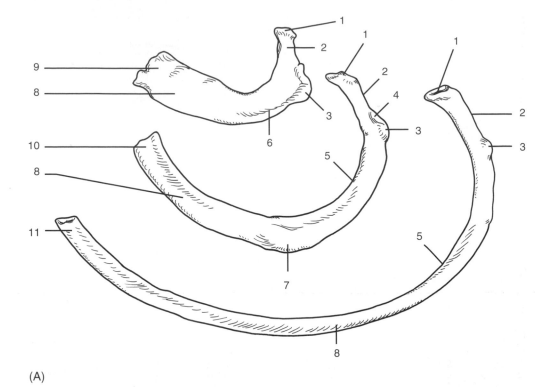

(A)

(B)

Figure 4.26   Ribs one, two, and typical rib—superior view.

1. Head of rib
2. Neck of rib
3. Tubercle of rib
4. Articular facet of tubercle
5. Costal angle
6. Tuberosity of scalenus medial muscle
7. Tuberosity of serratus anterior muscle
8. Body of rib
9. First rib
10. Second rib
11. Typical rib

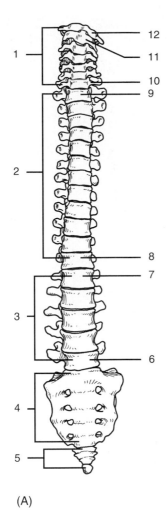

(A)

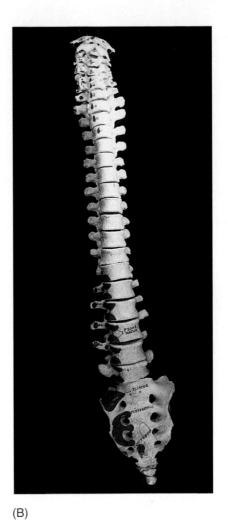

(B)

Figure 4.27   Vertebral column—anterior view.

1. Cervical vertebrae
2. Thoracic vertebrae
3. Lumbar vertebrae
4. Sacrum (S1–S5)
5. Coccyx
6. L5 vertebra

7. L1 vertebra
8. T12 vertebra
9. T1 vertebra
10. C7 vertebra
11. Axis C2 vertebra
12. Atlas C1 vertebra

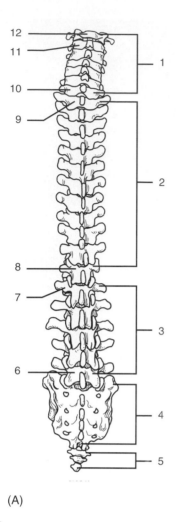

(A)

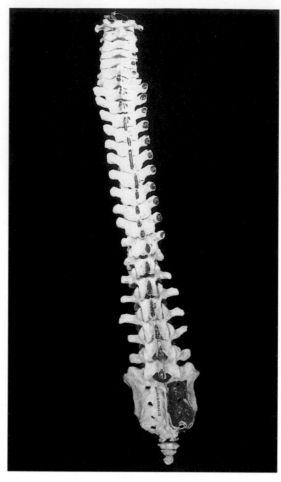

(B)

Figure 4.28   Vertebral column—posterior view.

1. Cervical vertebrae
2. Thoracic vertebrae
3. Lumbar vertebrae
4. Sacrum (S1–S5)
5. Coccyx
6. L5 vertebra

7. L1 vertebra
8. T12 vertebra
9. T1 vertebra
10. C7 vertebra
11. Axis C2 vertebra
12. Atlas C1 vertebra

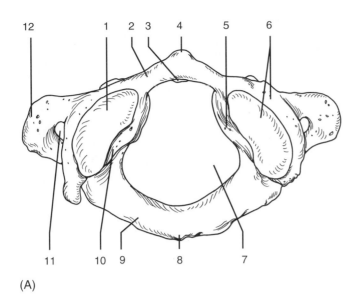

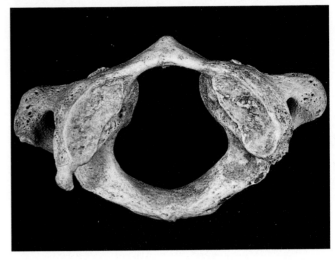

(A)

(B)

Figure 4.29 Atlas: first cervical vertebra—superior view.

1. Superior articular facet
2. Anterior arch
3. Articular facet for dens
4. Anterior tubercle
5. Tubercle for transverse ligament
6. Lateral mass

7. Vertebral foramen
8. Posterior tubercle
9. Posterior arch
10. Groove for vertebral artery
11. Transverse foramen
12. Transverse process

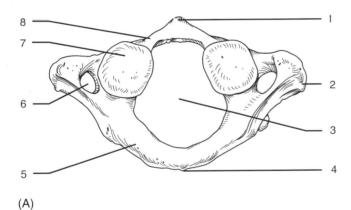

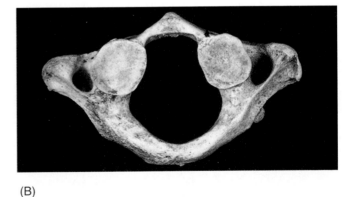

(A)

(B)

Figure 4.30 Atlas: first cervical vertebra—inferior view.

1. Anterior tubercle
2. Transverse process
3. Vertebral foramen
4. Posterior tubercle

5. Posterior arch
6. Transverse foramen
7. Inferior facet for axis
8. Anterior arch

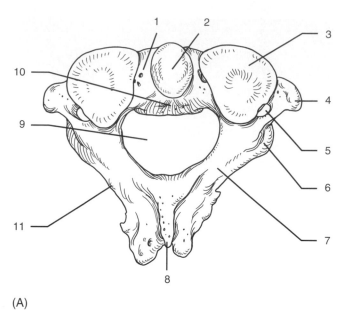

(A)

(B)

Figure 4.31   Axis: second cervical vertebra—superior view.

1. Pedicle
2. Dens (odontoid process)
3. Superior articular process (C1)
4. Transverse process
5. Transversarium foramen
6. Inferior articular process (C3)

7. Lamina
8. Spinous process (bifid)
9. Vertebral foramen
10. Body of axis
11. Arch of axis

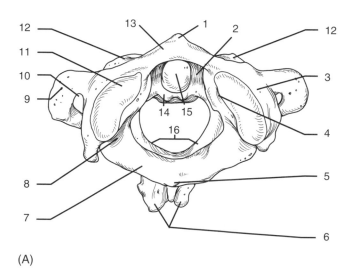

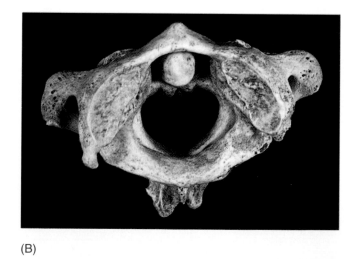

(A)

(B)

Figure 4.32   Atlas and axis articulated: first and second cervical vertebra—superior view.

1. Anterior tubercle of atlas
2. Pedicle of axis
3. Lateral mass of atlas
4. Tubercle for transverse ligament of atlas
5. Posterior tubercle of atlas
6. Spinous process (bifid) of axis
7. Posterior arch of atlas
8. Groove for vertebral artery of atlas

9. Transverse process of atlas
10. Transverse foramen of atlas
11. Superior articular facet of atlas
12. Superior articular process of axis
13. Anterior arch of atlas
14. Body of axis
15. Dens (odontoid process) of axis
16. Vertebral arch of axis

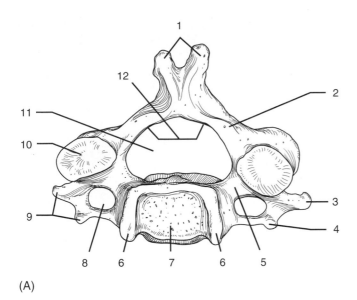

(A)

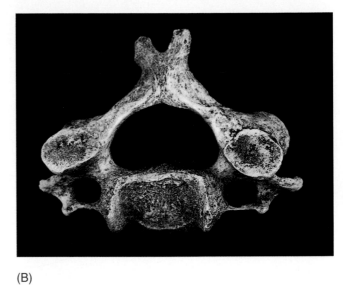

(B)

Figure 4.33   Typical cervical vertebra—superior view.

1. Bifid spinous process
2. Lamina
3. Posterior tubercle
4. Anterior tubercle
5. Pedicle
6. Uncus of vertebral body

7. Body of vertebra (centrum)
8. Transverse foramen
9. Transverse process
10. Superior articular facet
11. Vertebral foramen
12. Vertebral arch

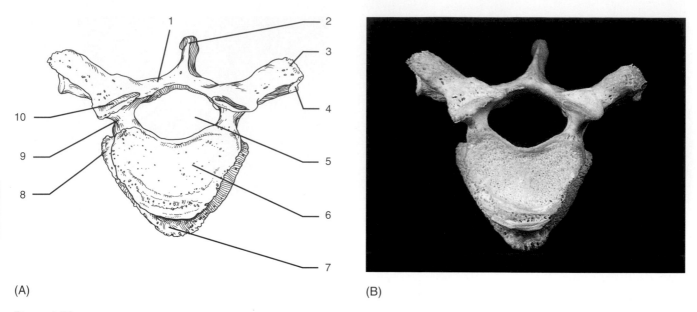

(A)

(B)

Figure 4.34 Typical thoracic vertebra. (a) and (b), superior view; (c) and (d), lateral view.

1. Lamina
2. Spinous process
3. Transverse process
4. Transverse costal facet
5. Vertebral foramen

6. Body
7. Annular apophysis
8. Superior costal facet
9. Pedicle
10. Superior articular process

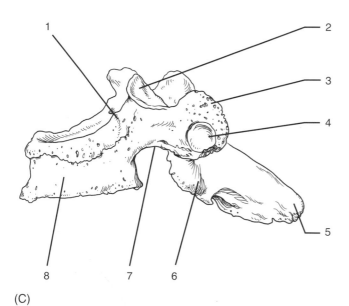

(C)

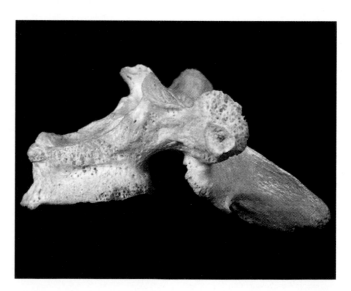

(D)

1. Superior costal facet
2. Superior articular process
3. Transverse process
4. Transverse costal facet

5. Spinous process
6. Inferior articular process
7. Inferior vertebral notch
8. Body

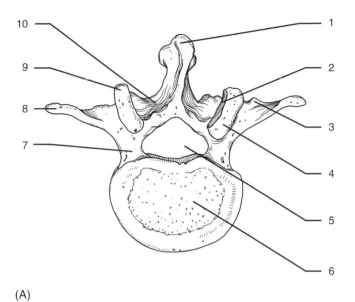

(A)

(B)

Figure 4.35 Lumbar vertebra—superior view.

1. Spinous process
2. Superior articular facet
3. Accessory process
4. Superior articular process
5. Vertebral foramen
6. Body of vertebra
7. Pedicle
8. Transverse process
9. Mamillary process
10. Lamina (vertebral arch)

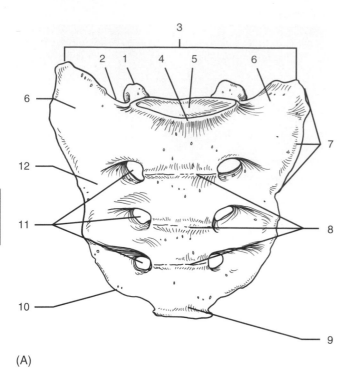

(A)

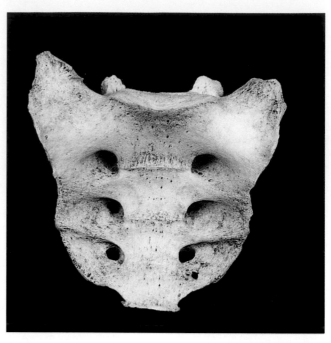

(B)

Figure 4.36   Sacrum—anterior view.

1. Superior articular process
2. Superior sacral notch
3. Base of sacrum
4. Sacral promontory
5. Lumbosacral articular surface
6. Sacral wings (ala)

7. Lateral part
8. Transverse lines (ridges)
9. Apex of sacrum
10. Inferolateral angle
11. Anterior sacral (pelvic) foramina
12. Lateral mass

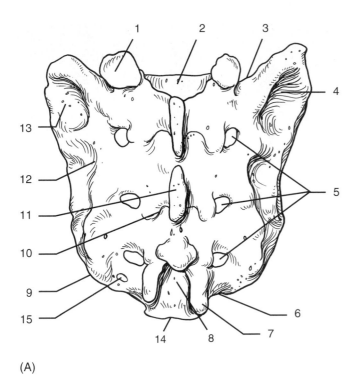

(A)

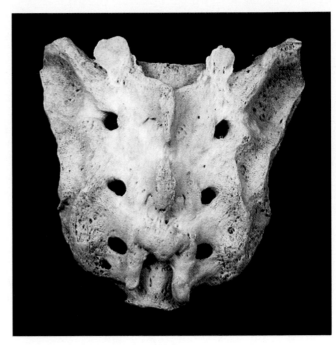

(B)

Figure 4.37 Sacrum—posterior view.

1. Superior articular process
2. Sacral canal
3. Superior sacral notch
4. Sacral tuberosity
5. Posterior (dorsal) sacral foramina
6. Sacrococcygeal notch
7. Sacral horn
8. Sacral hiatus

9. Inferolateral angle
10. Intermediate sacral crest
11. Median sacral crest
12. Lateral sacral crest
13. Auricular surface
14. Apex of sacrum
15. Underdeveloped foramen

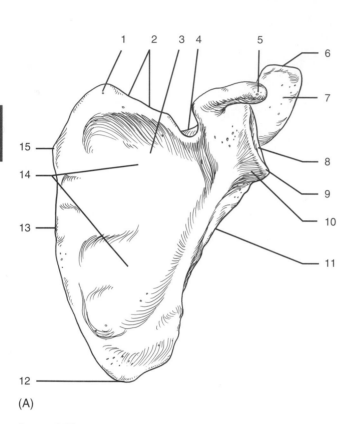

(A)

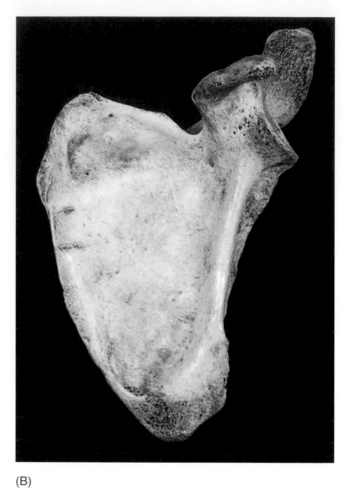

(B)

Figure 4.38   Left scapula. (a) and (b), anterior view; (c) and (d), posterior view.

1. Superior angle
2. Superior border
3. Subscapular fossa
4. Scapular notch
5. Coracoid process
6. Acromial articular surface
7. Acromion
8. Glenoid cavity
9. Infraglenoid tubercle
10. Neck of scapula
11. Lateral border
12. Inferior angle
13. Medial border
14. Costal surface
15. Medial angle

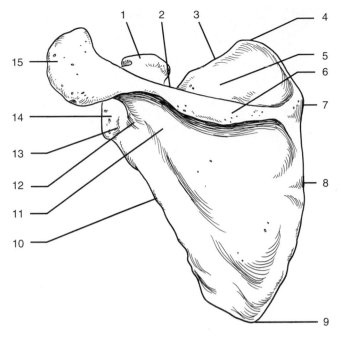

(C)

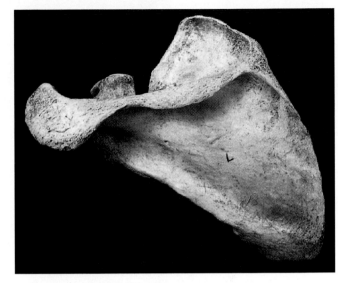

(D)

1. Coracoid process
2. Scapular notch
3. Superior border
4. Superior angle
5. Supraspinatus fossa
6. Spine of scapula
7. Medial angle
8. Medial border

9. Inferior angle
10. Lateral border
11. Infraspinatus fossa
12. Neck of scapula
13. Infraglenoid tubercle
14. Glenoid cavity
15. Acromion

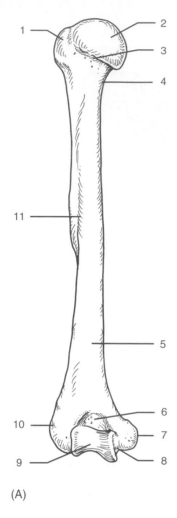

(A)

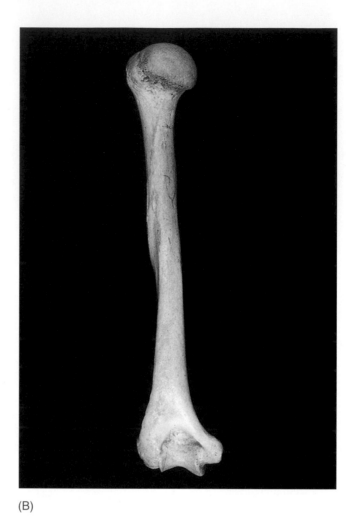

(B)

Figure 4.39   Left humerus. (a) and b, posterior view; (c) and (d), anterior view.

1. Greater tubercle
2. Head of humerus
3. Anatomical neck
4. Surgical neck
5. Posterior surface
6. Olecranon fossa

7. Medial epicondyle
8. Sulcus for ulnar nerve
9. Trochlea
10. Lateral epicondyle
11. Sulcus for radial nerve

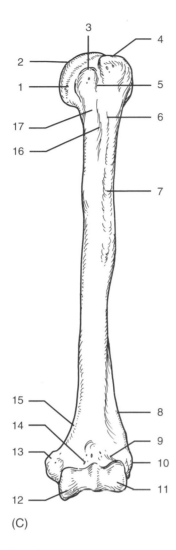

(C)

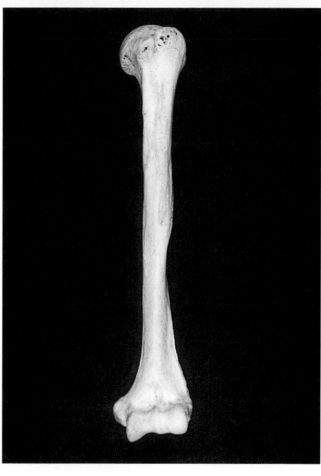

(D)

1. Anatomical neck
2. Head of humerus
3. Lesser tubercle
4. Greater tubercle
5. Intertubercular sulcus
6. Crest of greater tubercle
7. Deltoid tuberosity
8. Lateral supracondylar ridge
9. Radial fossa

10. Lateral epicondyle
11. Capitulum
12. Trochlea
13. Medial epicondyle
14. Coronoid fossa
15. Medial supracondylar ridge
16. Crest of lesser tubercle
17. Surgical neck

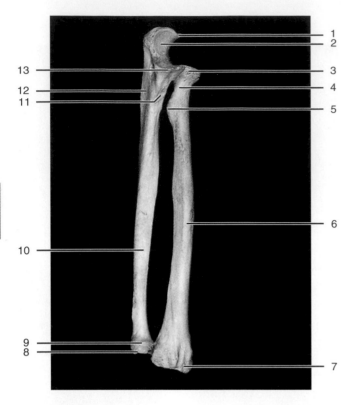

Figure 4.40   Left radius and ulna in anterior position.

1. Olecranon process
2. Trochlear notch
3. Head of radius
4. Neck of radius
5. Radial tuberosity
6. Anterior border
7. Styloid process of radius
8. Styloid process of ulna
9. Articular circumference
10. Anterior border of ulna
11. Ulnar tuberosity
12. Supinator crest
13. Coronal process

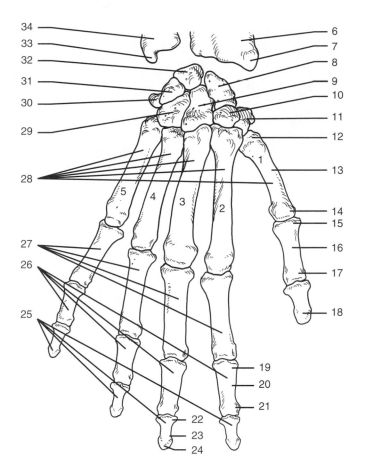

(A)

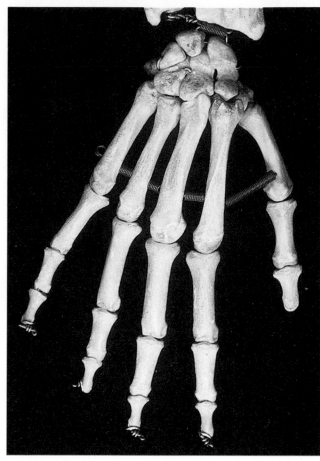

(B)

Figure 4.41   Right hand. (a) and (b), dorsal aspect; (c) and (d), palmar aspect.

1. First metacarpal
2. Second metacarpal
3. Third metacarpal
4. Fourth metacarpal
5. Fifth metacarpal
6. Radius
7. Styloid process of radius
8. Scaphoid (carpal) bone
9. Capitate (carpal) bone
10. Trapezoid (carpal) bone
11. Trapezium (carpal) bone
12. Base of metacarpal
13. Body of metacarpal
14. Head of metacarpal
15. Base of proximal phalanx
16. Body of proximal phalanx
17. Head of proximal phalanx

18. Thumb
19. Base of middle phalanx
20. Body of middle phalanx
21. Head of middle phalanx
22. Base of distal phalanx
23. Tuberosity of distal phalanx
24. Head of distal phalanx
25. Distal phalanges
26. Middle phalanges
27. Proximal phalanges
28. Metacarpal bones
29. Hamate (carpal) bone
30. Pisiform (carpal) bone
31. Triquetral (carpal) bone
32. Lunate (carpal) bone
33. Styloid process of ulna
34. Ulna

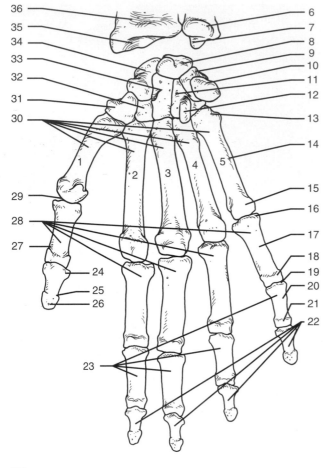

(C)

(D)

Figure 4.41   (Continued)

1. First metacarpal
2. Second metacarpal
3. Third metacarpal
4. Fourth metacarpal
5. Fifth metacarpal
6. Ulna
7. Styloid process of ulna
8. Lunate (carpal) bone
9. Triquetral (carpal) bone
10. Pisiform (carpal) bone
11. Hamate (carpal) bone
12. Hamulus (hook) of hamate bone
13. Base of metacarpal bone
14. Body of metacarpal bone
15. Head of metacarpal bone
16. Base of proximal phalanx
17. Body of proximal phalanx
18. Head of proximal phalanx

19. Base of middle phalanx
20. Body of middle phalanx
21. Head of middle phalanx
22. Distal phalanges
23. Middle phalanges
24. Base of distal phalanx
25. Tuberosity of distal phalanx
26. Head of distal phalanx
27. Thumb
28. Proximal phalanges
29. Interphangeal joint
30. Metacarpal bones
31. Trapezium (carpal) bone
32. Trapezoid (carpal) bone
33. Capitate (carpal) bone
34. Scaphoid bone
35. Styloid process of radius
36. Radius

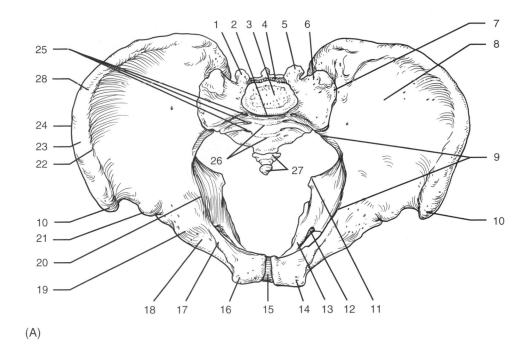

(A)

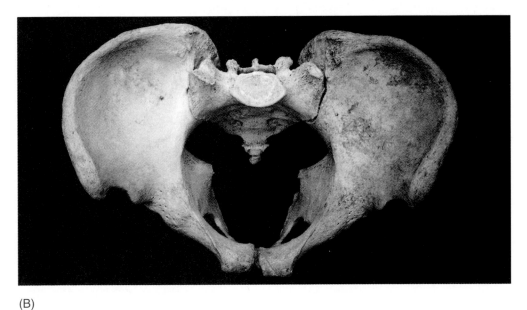

(B)

Figure 4.42   Pelvic girdle—superior view.

1. Superior articular process of sacrum
2. Sacral promontory
3. Base of sacrum
4. Sacral canal
5. Superior articular process of sacrum
6. Sacral wings (ala)
7. Sacroiliac joint
8. Iliac fossa
9. Linea terminalis
10. Anterior superior iliac spine
11. Ischial spine
12. Obturator foramen
13. Ischial tuberosity
14. Pubic crest

15. Pubic symphysis
16. Pubic tubercle
17. Pecten of pubis
18. Superior pubic ramus
19. Iliopubic eminence
20. Arcuate line
21. Anterior inferior iliac spine
22. Internal lip of iliac crest
23. Intermediate line of iliac crest
24. External lip of iliac crest
25. Anterior sacral (pelvic) foramina
26. Transverse lines (ridges) of sacrum
27. Coccyx
28. Iliac crest

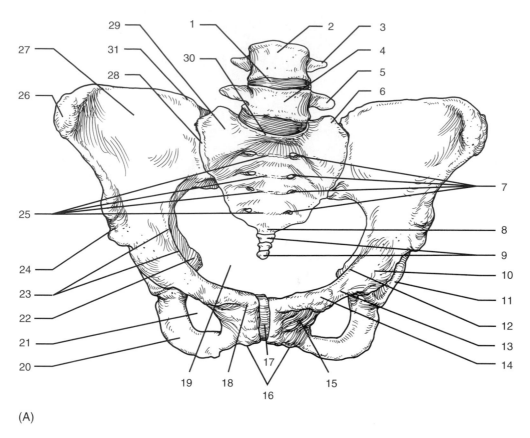

(A)

Figure 4.43 Female pelvic girdle—anterosuperior view.

1. Intervertebral disc
2. Body of L4 vertebra
3. Transverse process of L4 vertebra
4. Body of L5 vertebra
5. Transverse process of L5 vertebra
6. Base of sacrum
7. Anterior sacral (pelvic) foramina
8. Apex of sacrum
9. Coccyx
10. Iliopubic eminence
11. Acetabular fossa
12. Pecten of pubis
13. Superior pubic ramus
14. Pubic tubercle
15. Inferior pubic ramus
16. Pubic arch
17. Pubic symphysis
18. Pubic crest
19. Inlet of pubis
20. Ischial tuberosity
21. Obturator foramen
22. Ischial spine
23. Linea terminalis
24. Anterior superior iliac spine
25. Transverse lines (ridges) of sacrum
26. Iliac crest
27. Iliac fossa
28. Sacroiliac joint
29. Sacral wings (ala)
30. Sacral promontory
31. Iliac tuberosity

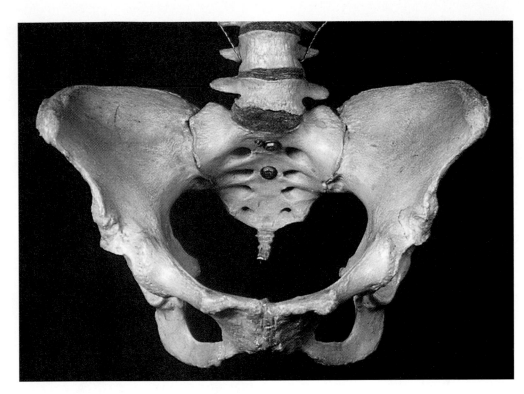

(B)

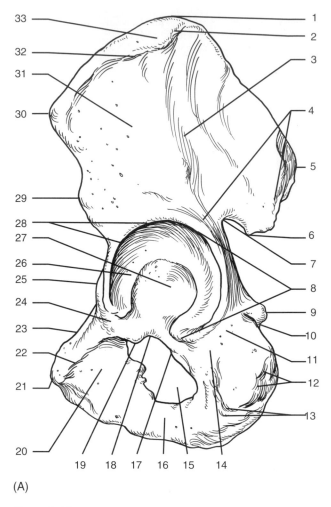

(A)

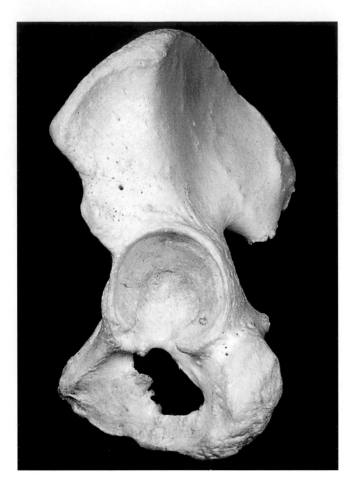

(B)

Figure 4.44   Os coxae—lateral view.

1. Iliac crest
2. Intermediate line of iliac crest
3. Anterior gluteal line
4. Inferior gluteal line
5. Posterosuperior iliac spine
6. Posteroinferior iliac spine
7. Greater sciatic notch
8. Acetabulum
9. Ischial spine
10. Lesser sciatic notch
11. Body of ischium
12. Ischial tuberosity
13. Ramus of ischium
14. Ischium
15. Obturator foramen
16. Inferior ramus of pubis
17. Posterior obturator tubercle

18. Acetabular notch
19. Anterior obturator tubercle
20. Body of pubis
21. Pubic tubercle
22. Obturator crest
23. Pubic crest
24. Body of ischium
25. Iliopubic eminence
26. Lunate surface of acetabulum
27. Acetabular fossa
28. Rim of acetabulum
29. Anterior inferior iliac spine
30. Anterior superior iliac spine
31. Ilium
32. External lip of iliac crest
33. Tuberosity of ilium

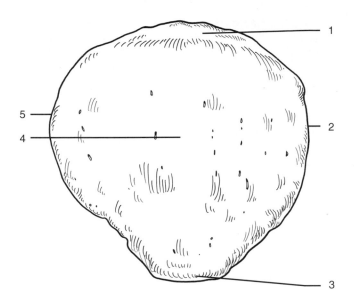

(A)

(B)

Figure 4.45   Left patella—anterior view.

1. Base (superior border)
2. Lateral border
3. Apex

4. Anterior surface of patella
5. Medial border

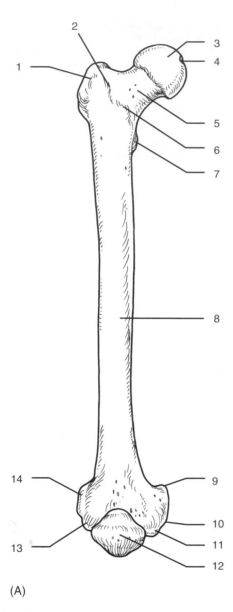

(A)

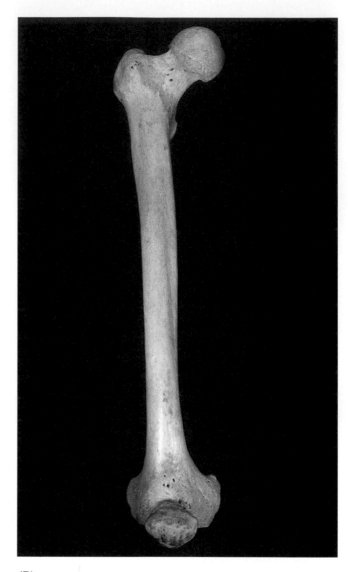

(B)

Figure 4.46   Right femur—anterior view.

1. Greater trochanter
2. Quadrate tubercle
3. Head
4. Fovea capitis femoris
5. Neck
6. Intertrochanteric line
7. Lesser trochanter

8. Shaft of femur
9. Adductor tubercle
10. Medial epicondyle
11. Medial condyle
12. Anterior surface of patella
13. Lateral condyle
14. Lateral epicondyle

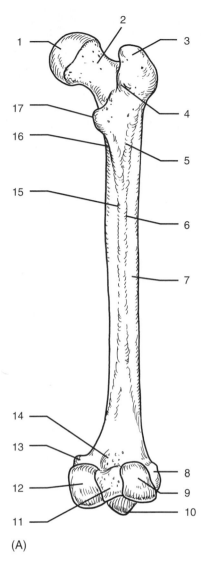

(A)

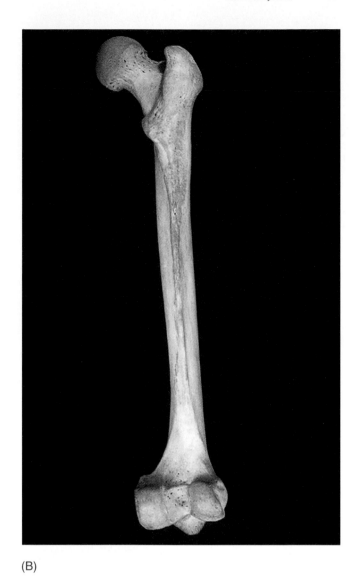

(B)

Figure 4.47   Right femur—posterior view.

1. Head
2. Neck
3. Greater trochanter
4. Intertrochanteric crest
5. Gluteal tuberosity
6. Lateral lip of linea aspera
7. Shaft (diaphysis)
8. Lateral epicondyle
9. Lateral condyle

10. Apex of patella
11. Intercondylar fossa
12. Medial condyle
13. Adductor tubercle
14. Popliteal surface
15. Medial lip of linea aspera
16. Pectineal line
17. Lesser trochanter

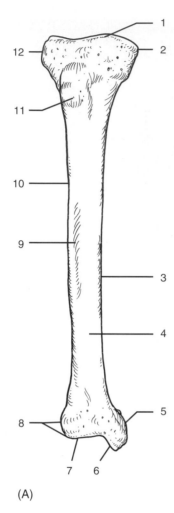

(A)

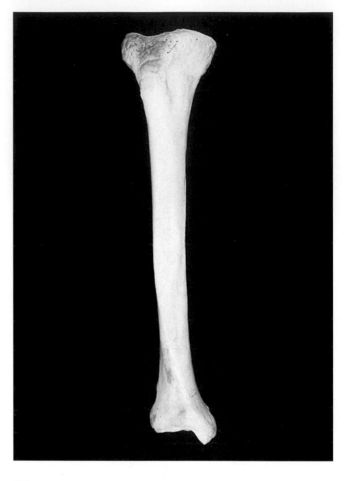

(B)

Figure 4.48   Right tibia—anterior view.

1. Anterior intercondylar area
2. Medial condyle
3. Medial border
4. Shaft of tibia (diaphysis)
5. Medial malleolus
6. Malleolar articular surface
7. Inferior articular surface
8. Fibular notch
9. Anterior border (crest)
10. Interosseous border
11. Tuberosity of tibia
12. Lateral condyle

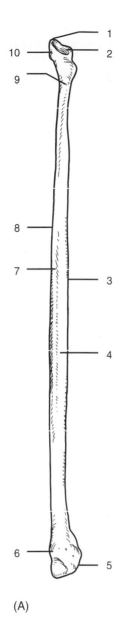

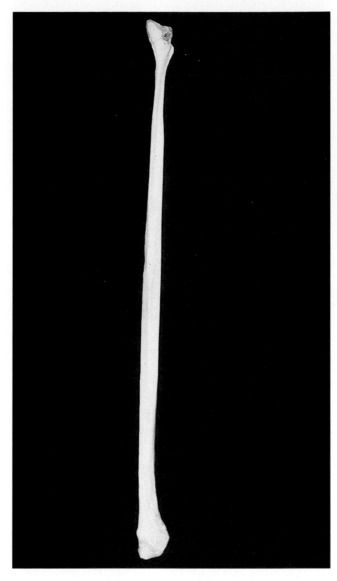

(A)   (B)

Figure 4.49   Right fibula—anterior view.

1. Apex
2. Position of tibiofibular joint
3. Interosseous border
4. Fibula
5. Malleolar articular surface

6. Lateral malleolus
7. Anterior border
8. Lateral border
9. Neck of fibula
10. Head

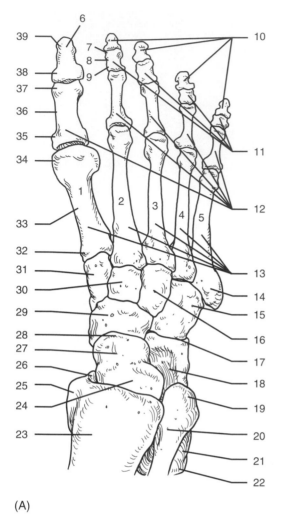

(A)

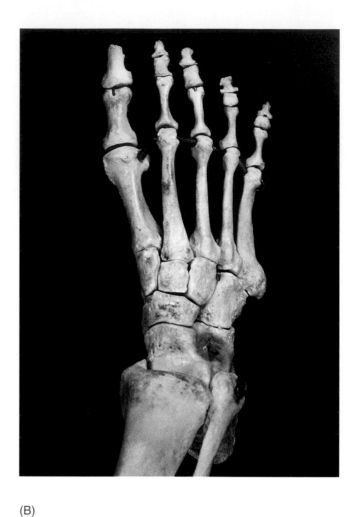

(B)

Figure 4.50   Right foot—dorsal aspect with distal end of tibia and fibula.

1. First metatarsal bone
2. Second metatarsal bone
3. Third metatarsal bone
4. Fourth metatarsal bone
5. Fifth metatarsal bone
6. Distal phalanx (great toe)
7. Head of middle phalanx
8. Body of middle phalanx
9. Base of middle phalanx
10. Distal phalanges
11. Middle phalanges
12. Proximal phalanges
13. Metatarsal bones
14. Tuberosity of fifth metatarsal bone
15. Cuboid bone
16. Lateral cuneiform bone
17. Calcaneocuboidal joint
18. Tarsal sinus
19. Lateral malleolus of fibula
20. Fibula
21. Calcaneous bone
22. Calcaneal tuberosity
23. Tibia bone
24. Trochlea of talus
25. Medial condyle of tibia
26. Medial malleolus of tibia
27. Neck of talus
28. Head of talus
29. Navicular bone
30. Intermediate cuneiform bone
31. Medial cuneiform bone
32. Base of metatarsal bone
33. Body of metatarsal bone
34. Head of metatarsal bone
35. Base of proximal phalanx
36. Body of proximal phalanx
37. Head of proximal phalanx
38. Base of distal phalanx
39. Tuberosity of distal phalanx

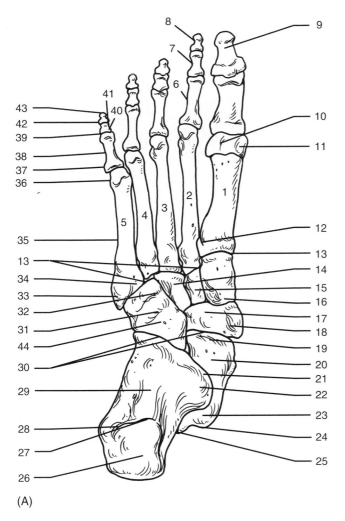

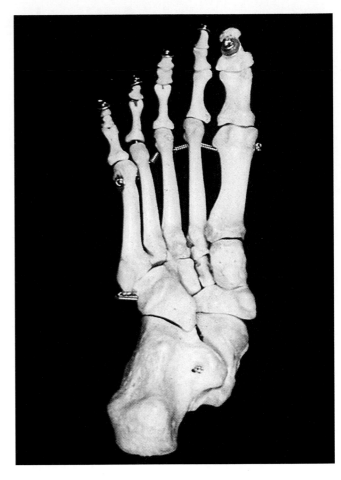

(A)

(B)

Figure 4.51 Right foot—inferior (plantar) view.

1. First metatarsal
2. Second metatarsal
3. Third metatarsal
4. Fourth metatarsal
5. Fifth metatarsal
6. Proximal phalanx
7. Middle phalanx
8. Distal phalanx
9. Great toe
10. Lateral sesamoid bone
11. Medial sesamoid bone
12. Tuberosity of first metatarsal
13. Tarsometatarsal joint
14. Lateral cuneiform bone
15. Intermediate cuneiform bone
16. Medial cuneiform bone
17. Navicular bone
18. Tuberosity of navicular bone
19. Head of talus
20. Talus
21. Sustentaculum tali
22. Groove for flexor hallucis longus tendon

23. Posterior process of talus
24. Medial tubercle of talus
25. Lateral tubercle of talus
26. Tuberosity of calcaneus
27. Medial process of calcaneus
28. Lateral process of calcaneus
29. Calcaneus
30. Transverse tarsal joint
31. Tuberosity of cuboid bone
32. Groove for peroneus longus tendon
33. Tuberosity of fifth metatarsal
34. Base of fifth metatarsal
35. Body of fifth metatarsal
36. Head of fifth metatarsal
37. Base of proximal phalanx
38. Body of proximal phalanx
39. Head of proximal phalanx
40. Base of middle phalanx
41. Head of middle phalanx
42. Base of distal phalanx
43. Tuberosity of distal phalanx
44. Cuboid bone

Arthrology (Joints)

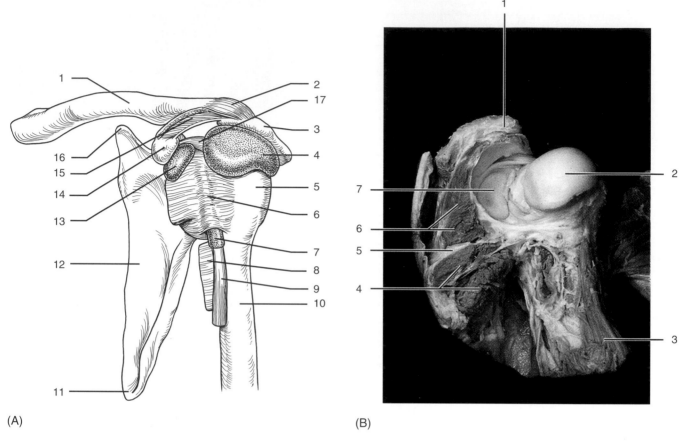

(A)   (B)

Figure 4.52   (a) Shoulder joint—anterior view. (b) Shoulder joint dissection with scapula reflected laterally.

1. Clavicle
2. Acromioclavicular ligament
3. Acromion
4. Subacromial bursa
5. Greater tubercle humerus
6. Transverse ligament of humerus
7. Intertubercular synovial sheath
8. Teres major muscle
9. Biceps brachii tendon (long head)
10. Humerus (diaphysis)
11. Inferior angle
12. Scapula
13. Subcoracoid bursa
14. Coracoid process
15. Coracoacromial ligament
16. Superior angle
17. Supraspinatus tendon

1. Acromion process of scapula
2. Head of humerus
3. Latissimus dorsi muscle
4. Infra spinatus muscle
5. Spine of scapula
6. Supra spinatus muscle
7. Glenoid fossa

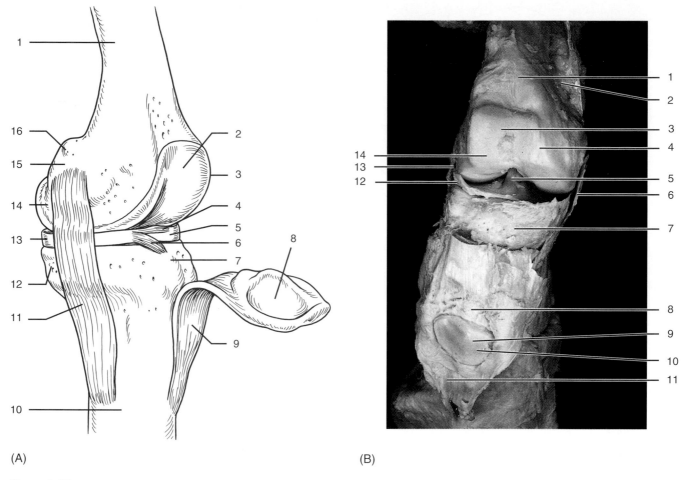

Figure 4.53 (a) Knee joint—lateral anterior view. (b) Anterior view of knee joint; patella with ligament reflected to show underlying structures.

1. Femur
2. Patellar articular surface
3. Lateral femoral condyle
4. Transverse ligament of knee
5. Lateral meniscus
6. Medial meniscus
7. Lateral condyle of tibia
8. Patella (articular surface)
9. Patellar ligament
10. Tibia
11. Tibial collateral ligament
12. Medial condyle of tibia
13. Medial meniscus
14. Articular surface of medial posterior femoral condyle
15. Medial epicondyle of femur
16. Adductor tubercle

1. Suprapatellar bursa
2. Capsule
3. Patellar surface
4. Medial condyle of femur
5. Anterior cruciate ligament
6. Medial collateral ligament
7. Infrapatellar fat pad
8. Ligamentum patellae
9. Patellar vertical ridge
10. Patellar articular facet
11. Tendon of rectus femoris muscle
12. Lateral meniscus
13. Lateral collateral ligament
14. Articular cartilage

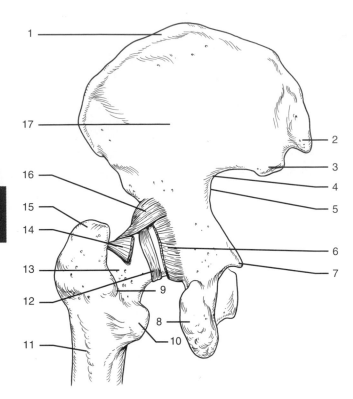

Figure 4.54 Left pelvis and hip joint—posterior view.

1. Iliac crest
2. Superior iliac spine
3. Inferior iliac spine
4. Greater sciatic notch
5. Pelvic brim
6. Iliofemoral ligament (cut)
7. Ischial spine
8. Ischial tuberosity
9. Intertrochanteric crest
10. Lesser trochanter
11. Femur
12. Pubofemoral ligament
13. Neck of femur
14. Ischiofemoral ligament (cut)
15. Greater trochanter
16. Iliofemoral ligament
17. Iliac fossa

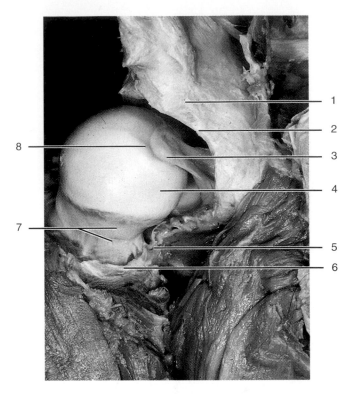

Figure 4.55 Right hip joint with femur head reflected laterally.

1. Iliopubic eminence
2. Acetabular labrum
3. Ligamentum teres of femur
4. Head of femur
5. Obturator externus muscle
6. Capsule reflected
7. Retinacular fibers
8. Fovea covered by ligamentum capitis femoris

Radiographs

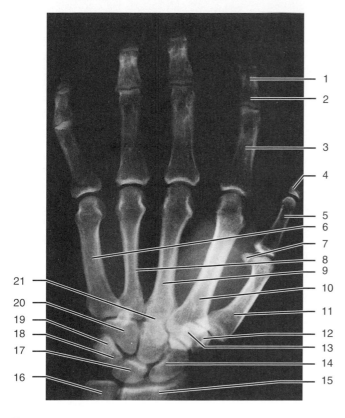

Figure 4.56 Radiograph of adult left hand.

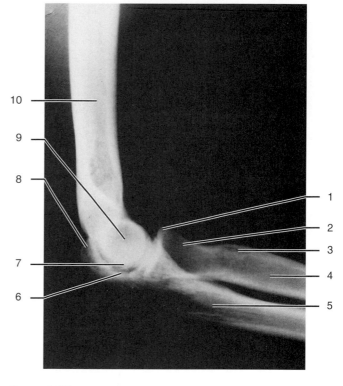

Figure 4.57 Radiograph of elbow joint in a right-angle flexion.

1. Distal phalanx of index finger
2. Middle phalanx of index finger
3. Proximal phalanx
4. Distal phalanx
5. Proximal phalanx
6. Fifth metacarpal
7. Sesamoid bone
8. Fourth metacarpal
9. Third metacarpal
10. Second metacarpal
11. First metacarpal
12. Trapezium
13. Trapezoid
14. Scaphoid
15. Radius
16. Ulna
17. Lunate
18. Triquetral
19. Pisiform
20. Hamate
21. Capitate

1. Head of radius
2. Neck of radius
3. Tuberosity of radius
4. Radius
5. Ulna
6. Trochlear notch of ulna
7. Trochlea of humerus
8. Olecranon process
9. Humeral condyle
10. Humerus

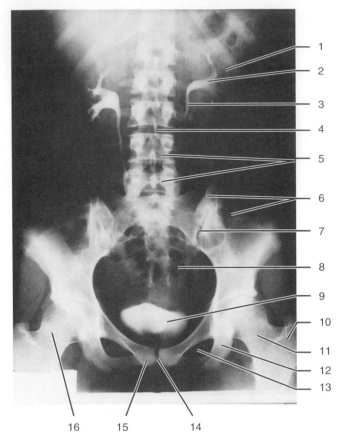

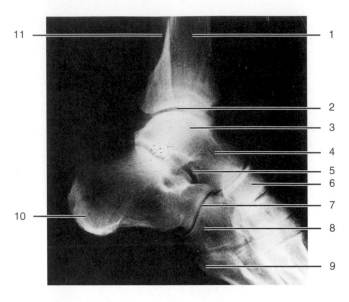

Figure 4.59 Radiograph of ankle—lateral aspect.

1. Tibia
2. Talotibial joint
3. Talus
4. Neck of talus
5. Tarsal sinus
6. Navicular
7. Calcaneocuboid joint
8. Cuboid
9. Fifth metatarsal tuberosity
10. Calcaneus
11. Fibula

Figure 4.58 Radiograph of anterior-posterior projection of pelvic and lumbar region.

1. Kidney
2. Pelvis of kidney
3. Ureter
4. Intervertebral disc
5. Lumbar vertebrae
6. Ilium
7. Sacroiliac joint
8. Sacrum
9. Urinary bladder
10. Greater trochanter of femur
11. Neck of femur
12. Ischial tuberosity
13. Obturator foramen
14. Pubic symphysis
15. Pubic tubercle
16. Head of femur

CHAPTER FIVE

Muscular System

B ased on its morphology and functions, muscle tissue can be classified as smooth muscle, skeletal muscle, or cardiac muscle. The smooth muscle cells are called smooth since they lack organized striations of actin and myosin proteins. Smooth muscle consists of involuntary fibers found in the body wall of the gastrointestinal system, in the gall bladder, urinary bladder and urinary tract, uterus, blood vessels, around exocrine glands, and in body walls of other hollow organs. Morphologically, the smooth muscle cells are short and spindle shaped, with a centrally located nucleus.

Skeletal muscle, or striated muscle fiber, is generally associated with bones. However, the striated fibers are also found in other parts of the body. The individual muscle cell is surrounded by a thin layer of connective tissue, the endomysium. The muscle cells are multinucleated, and display well-organized thick bands of myosin and thin bands of actin. The cells form a long cylindrical fiber that is tapered at both ends. Groups of muscle fibers in a muscle form a fascicle. The fasciculi are surrounded by connective tissue forming outer covering, the epimysium.

Connective tissue fascia separates individual muscles. The fascia can extend beyond the length of the muscle fibers to form a cordlike tendon. Tendons in turn attach to the periosteum of a bone. Connective tissue may also form an aponeurosis, a broad fibrous covering around the muscle. The aponeurosis of one muscle may connect to the adjoining muscle's aponeurosis.

Cardiac muscle fibers are confined to the heart. The fibers are striated, branched, and are joined end to end by intercalated discs. Within the cell, the sarcoplasm contains a centrally placed nucleus, dense concentrations of actin and myosin filaments, mitochondria, and well-developed sarcoplasmic reticulum with cisternae and transverse tubules that extend through the width of the fiber.

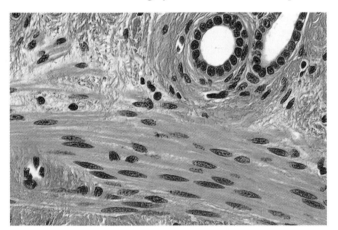

Figure 5.1   Light micrograph (LM) of smooth muscle. Also seen in the micrograph are two excretory ducts. (400×)

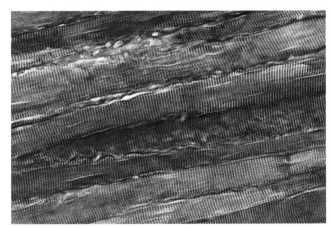

Figure 5.2   LM of skeletal muscle. (400×)

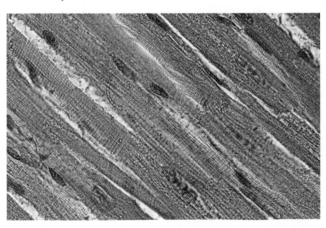

Figure 5.3   LM of cardiac muscle. (400×)

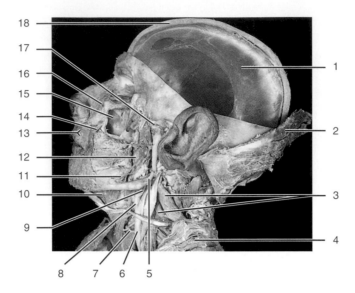

Figure 5.4   Head and neck in lateral view

1. Falx cerebri
2. Sternocleidomastoid (reflected)
3. Levator scapulae muscle
4. Trapezius muscle
5. Angle of mandible
6. Common carotid artery
7. Superior thyroid artery
8. External carotid artery
9. Ascending pharyngeal artery
10. Myohyoid muscle
11. Inferior alveolar artery and nerve
12. Medial pterygoid muscle
13. Orbicularis oris muscle
14. Zygomaticus minor muscle
15. Mucous membrane of maxillary sinus
16. Levator labii superious muscle
17. Temporomandibular joint
18. Diploë

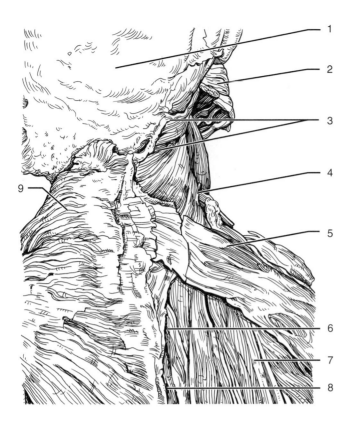

(A)

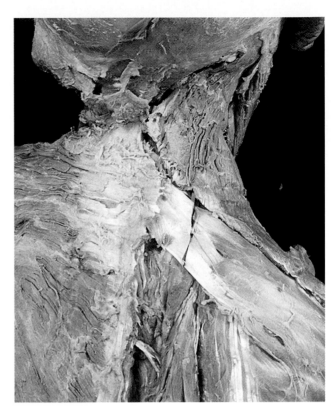

(B)

Figure 5.5 Muscles of neck and upper back—posterior view.

1. External occipital protuberance
2. Sternocleidomastoid
3. Splenius capitis muscle
4. Serratus posterior superior

5. Longissimus thoracis muscle
6. Tendons of iliocostalis thoracis
7. Longissimus cervicis muscle
8. Trapezius muscle

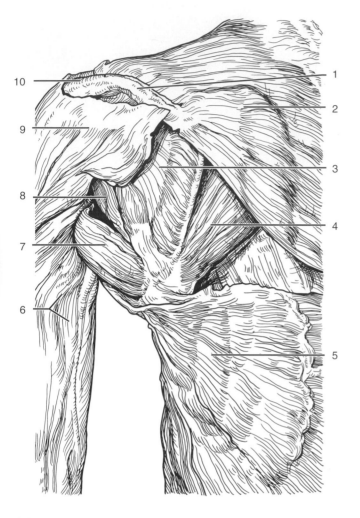

(A)

(B)

Figure 5.6 Muscles of upper back and arm.

1. Spine of scapula
2. Trapezius muscle
3. Infraspinatus muscle
4. Rhomboideus muscle
5. Latissimus dorsi muscle

6. Triceps brachii muscle
7. Teres major muscle
8. Teres minor muscle
9. Deltoid (posterior medial muscle scapular head)
10. Acromion process of scapula

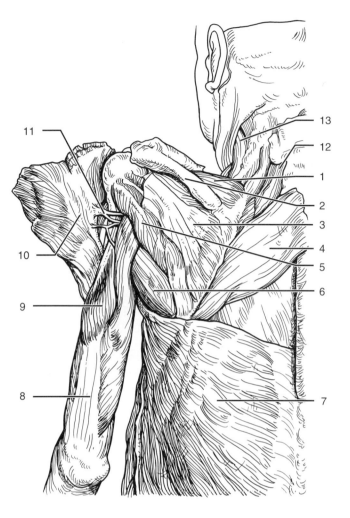

(A)

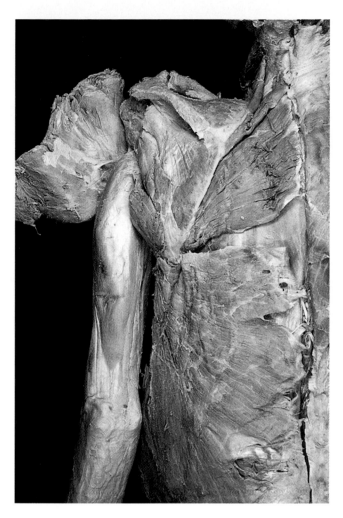

(B)

Figure 5.7 Muscles of shoulder with deltoid muscle reflected.

1. Supraspinatus muscle
2. Spine of scapula
3. Infraspinatus muscle
4. Rhomboid major muscle
5. Teres minor muscle
6. Teres major muscle
7. Latissimus dorsi muscle
8. Triceps brachii tendon
9. Lateral head of triceps brachii muscle
10. Deltoid muscle reflected
11. Axillary nerve and posterior circumflex humeral artery
12. Rhomboideus minor muscle reflected
13. Levator scapulae muscle

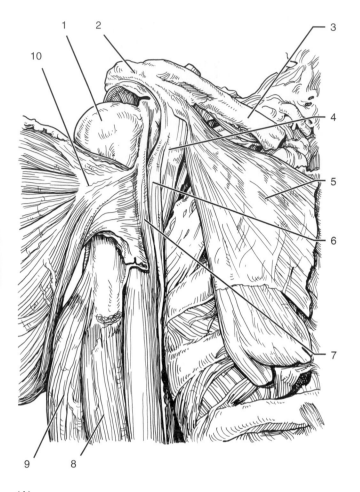

(A)

(B)

Figure 5.8   Right shoulder—anterior view.

1. Head of humerus
2. Acromion process
3. Midclavicle
4. Coracobrachialis muscle
5. Pectoralis minor muscle

6. Short head of biceps brachii muscle
7. Long head of biceps brachii muscle
8. Brachialis muscle
9. Lateral head of triceps brachialis muscle
10. Pectoralis major muscle

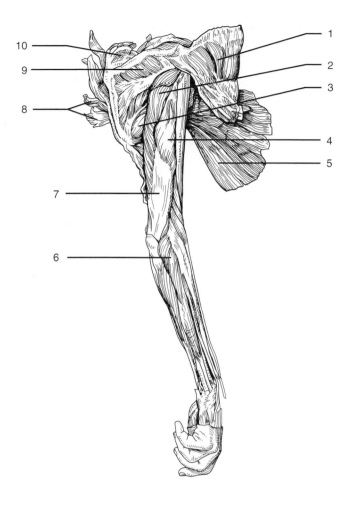

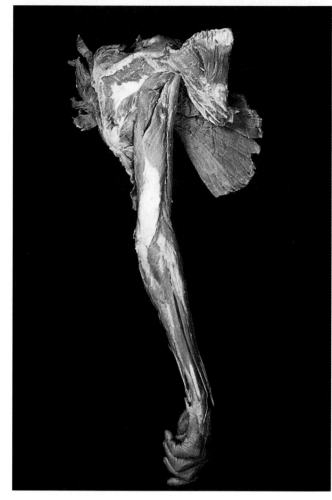

(A)

(B)

Figure 5.9   Muscles of the upper arm—posterior view.

1. Deltoid muscle
2. Triceps brachii, long head muscle
3. Teres major muscle
4. Triceps brachii muscle, lateral head
5. Pectoralis major muscle

6. Anconeus muscle
7. Tendon of triceps brachii
8. Serratus anterior muscle
9. Infraspinatus muscle
10. Supraspinatus muscle

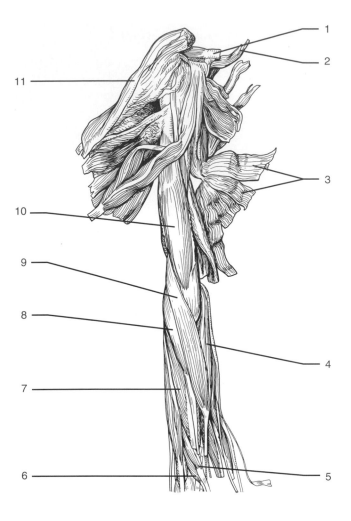

(A)

(B)

Figure 5.10 Muscle of upper arm—medial view.

1. Clavicle
2. Subclavius muscle
3. Serratus anterior muscle
4. Pronator teres muscle
5. Abductor pollicis longus muscle
6. Extensor pollicis brevis muscle

7. Extensor carpi radialis brevis muscle
8. Extensor carpi radialis longus muscle
9. Brachioradialis muscle
10. Biceps brachii muscle
11. Deltoid muscle

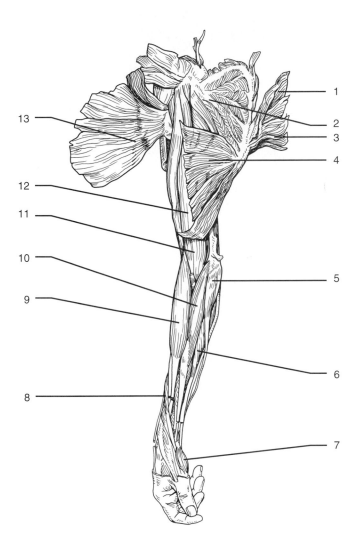

(A)

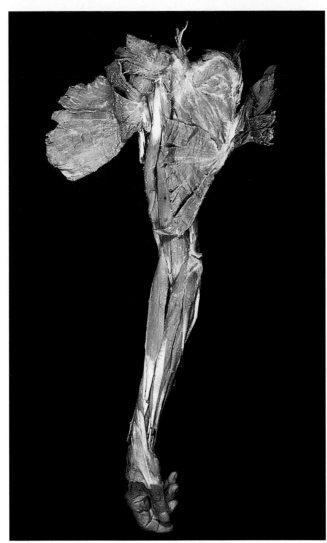

(B)

Figure 5.11   Deep muscles of the arm—anterior view

1. Rhomboideus minor muscle
2. Subscapularis muscle
3. Rhomboideus major muscle
4. Serratus anterior muscle
5. Pronator teres muscle
6. Flexor digitorum superficialis muscle
7. Abductor pollicis brevis muscle

8. Abductor pollicis longus muscle
9. Extensor carpi radialis longus muscle
10. Flexor carpi radialis muscle
11. Brachialis muscle
12. Biceps brachii muscle
13. Pectoralis major muscle

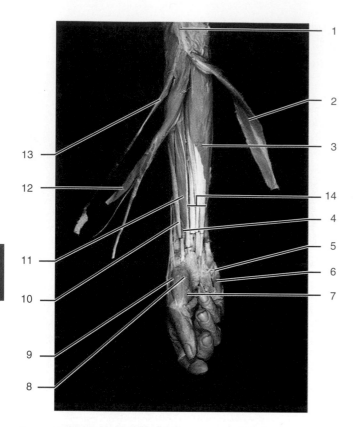

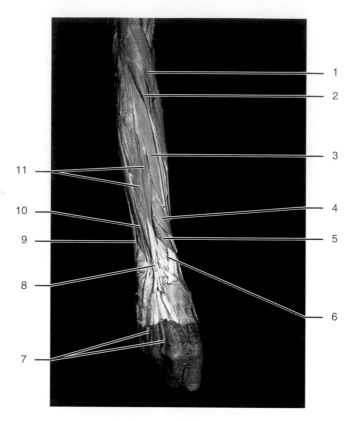

Figure 5.12   Muscles and tendons of the right arm—ventral aspect.

1. Medial epicondyle
2. Flexor carpi ulnaris muscle
3. Flexor digitorum profundus muscle
4. Median nerve
5. Flexor digiti minimi brevis muscle
6. Abductor digiti minimus muscle
7. Abductor pollicis muscle
8. Flexor pollicis brevis muscle
9. Abductor pollicis brevis muscle
10. Tendon of flexor carpi radialis
11. Flexor of pollicis longus muscle
12. Flexor digitorum superficialis muscle
13. Palmaris longus muscle
14. Tendon of flexor digitorum profundus muscle

Figure 5.13   Superficial muscles of the forearm—posterior view.

1. Brachioradialis muscle
2. Extensor carpi radialis longus muscle
3. Extensor carpi radialis brevis muscle
4. Abductor pollicis longus muscle
5. Extensor pollicis brevis muscle
6. Cephalic vein
7. Dorsal metacarpal veins
8. Basilic vein
9. Extensor carpi ulnaris muscle
10. Tensor digiti minimus muscle
11. Extensor digitorum muscle

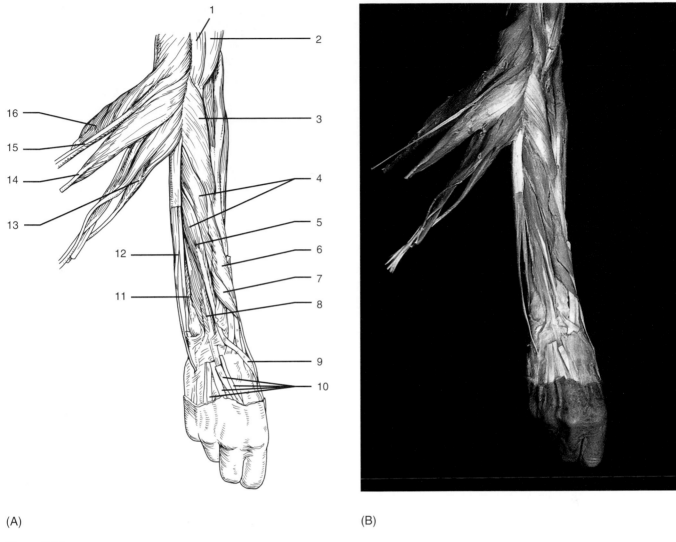

(A)   (B)

Figure 5.14   Muscles and tendons of the right forearm—dorsal view.

1. Brachialis muscle
2. Biceps brachii muscle
3. Supinator muscle
4. Extensor digitorum muscle
5. Extensor pollicis longus muscle
6. Abductor pollicis longus muscle
7. Extensor pollicis brevis
8. Extensor indicis muscle
9. Tendon of extensor pollicis longus muscle
10. Tendons of extensor digitorum muscle
11. Extensor carpi ulnaris muscle
12. Extensor digiti minimi muscle
13. Extensor digitorum muscle
14. Extensor carpi radialis brevis muscle
15. Extensor carpi radialis longus muscle
16. Brachioradialis muscle

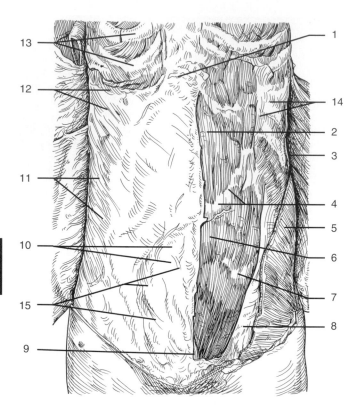

(A)

(B)

Figure 5.15 Superficial abdominal muscles—anterior view.

1. Level of xiphoid process of sternum
2. Rectus abdominis muscle
3. External oblique muscle
4. Tendinous intersection muscle
5. Internal oblique muscle
6. Rectus abdominis muscle
7. Tendinous intersection muscle
8. Transversus abdominis

9. Pyramidalis anterior muscle
10. Rectus sheath
11. External oblique muscle
12. Serratus anterior muscle
13. Ribs
14. External intercostal muscles
15. Transversalis fascia

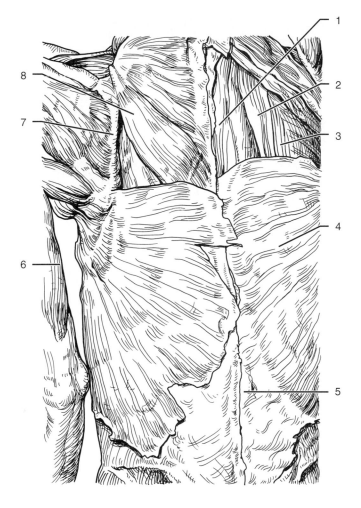

(A)

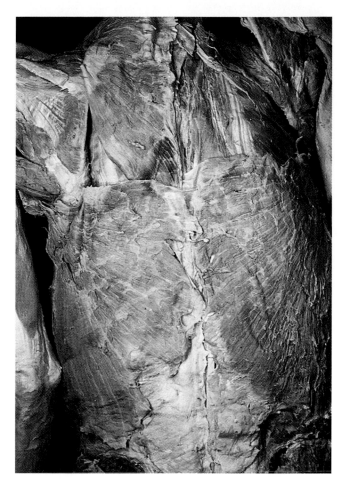

(B)

Figure 5.16 Muscles of the back—posterior view.

1. Spinalis thoracis muscle
2. Longissimus thoracis muscle
3. Tendon of iliocostalis thoracis
4. Latissimus dorsi muscle
5. Thoracolumbar fascia
6. Long head of triceps brachii muscle
7. Infraspinatus muscle
8. Trapezius muscle

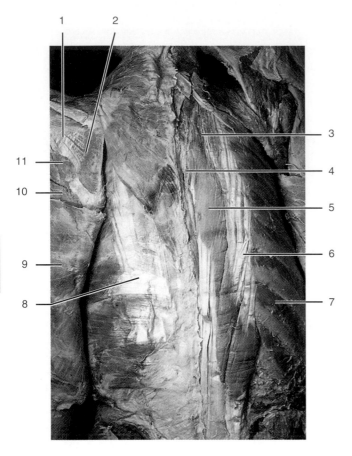

Figure 5.17   Deep muscles of the back.

1. Spine of scapula
2. Infraspinatus muscle
3. Iliocostalis cervicis muscle
4. Longissimus cervicis muscle
5. Longissimus thoracis muscle
6. Iliocostalis thoracis muscle
7. External intercostal muscle
8. Tendon of serratus posterior inferior
9. Latissimus dorsi muscle
10. Teres major muscle
11. Teres minor muscle

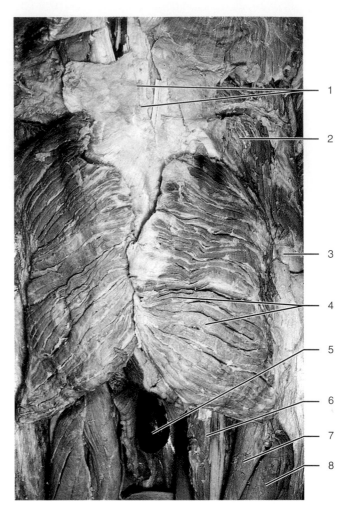

Figure 5.18   Superficial muscles of the hip—posterior view.

1. Thoracolumbar fascia
2. Gluteus medius muscle
3. Tensor fascia latae
4. Gluteus maximus muscle
5. Scrotum
6. Adductor magnus muscle
7. Semitendinosus muscle
8. Biceps femoris muscle (long head)

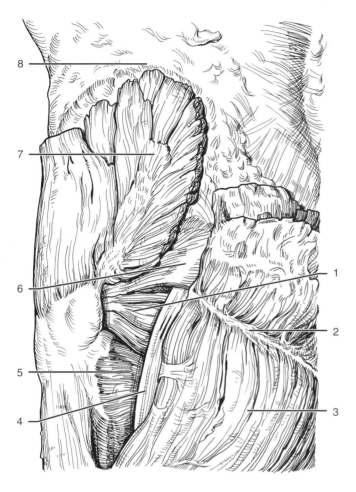

(A)

(B)

Figure 5.19   Superficial gluteal muscles.

1. Superior gemellus muscle
2. Inferior gluteal nerve
3. Gluteus maximus muscle (reflected)
4. Sciatic nerve

5. Quadratus femoris muscle
6. Piriformis medius muscle
7. Gluteus medius muscle
8. Iliac crest

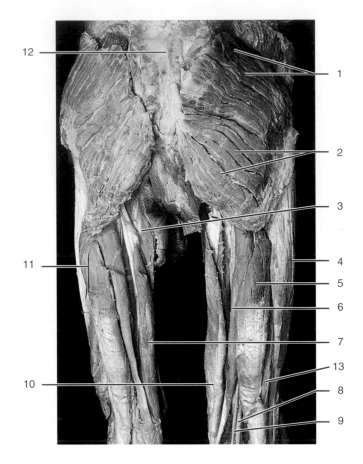

Figure 5.20 Superficial muscles of the thigh—posterior view.

1. Gluteus medius muscle
2. Gluteus maximus muscle
3. Adductor magnus muscle
4. Iliotibial tract
5. Biceps femoris muscle (long head)
6. Semitendinosus muscle
7. Semimembranosus muscle
8. Popliteal artery
9. Tibial nerve
10. Semimembranosus muscle
11. Vastus lateralis at origin
12. Lumbodorsal fascia
13. Biceps femoris muscle (short head)

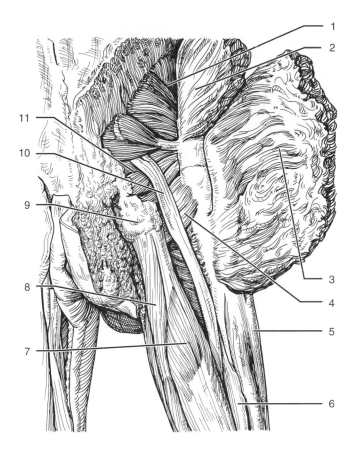

(A)

(B)

Figure 5.21   Deep gluteal muscles.

1. Gluteus minimus muscle
2. Gluteus medius muscle (reflected)
3. Gluteus maximus muscle (reflected)
4. Quadratus femoris muscle
5. Iliotibial tract
6. Biceps femoris muscle (short head)

7. Biceps femoris muscle (long head)
8. Semitendinosus muscle
9. Ischial tuberosity
10. Sciatic nerve
11. Piriformis muscle

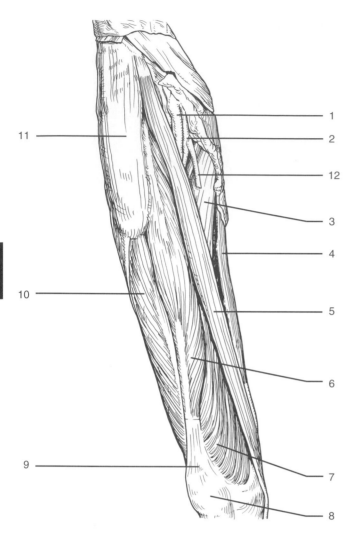

(A)

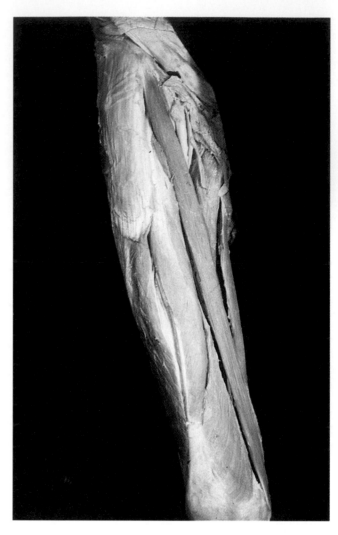

(B)

Figure 5.22 Superficial muscles of the upper leg—medial view.

1. Femoral artery
2. Femoral vein
3. Adductor longus muscle
4. Gracilis muscle
5. Sartorius muscle
6. Rectus femoris muscle
7. Vastus medialis muscle
8. Patella
9. Quadriceps femoris tendon
10. Vastus lateralis muscle
11. Tensor fasciae latae muscle
12. Great saphenous vein (cut)

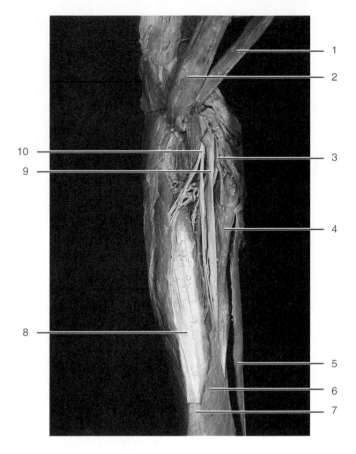

Figure 5.23 Blood vessels, nerves, and muscles of the thigh.

1. Sartorius muscle (reflected)
2. Rectus femoris muscle
3. Femoral vein
4. Adductor magnus muscle
5. Gracilis muscle
6. Vastus medialis muscle
7. Quadriceps femoris tendon (cut)
8. Vastus intermedius muscle
9. Femoral artery
10. Femoral nerve

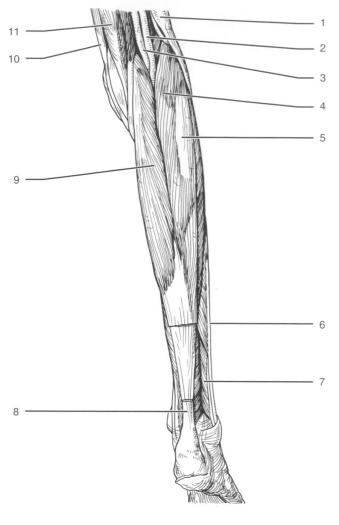

(A)

(B)

Figure 5.24   Muscles of the lower leg—posterior view.

1. Biceps femoris muscle
2. Popliteal artery
3. Tibial nerve
4. Plantaris muscle
5. Gastrocnemius muscle (lateral head)
6. Peroneus longus muscle

7. Peroneus brevis muscle
8. Calcaneal tendon
9. Gastrocnemius muscle (medial belly)
10. Gracilis muscle
11. Semimembranosus muscle

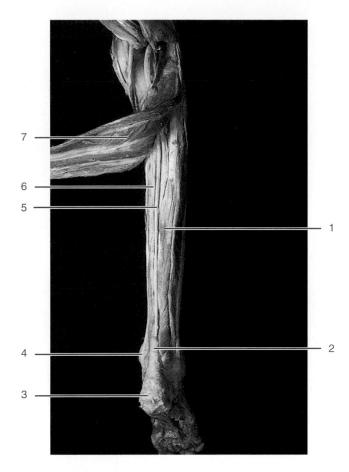

Figure 5.25   Muscles of the posterior crural region.

1. Tibialis posterior muscle
2. Calcaneal tendon (cut)
3. Calcaneal bone
4. Medial malleolus

5. Tibial nerve
6. Posterior tibial artery
7. Soleus muscle (reflected)

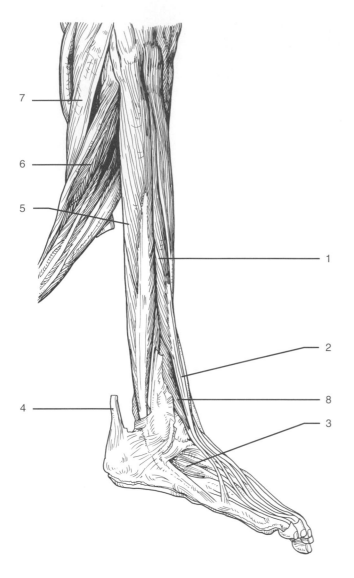

(A)

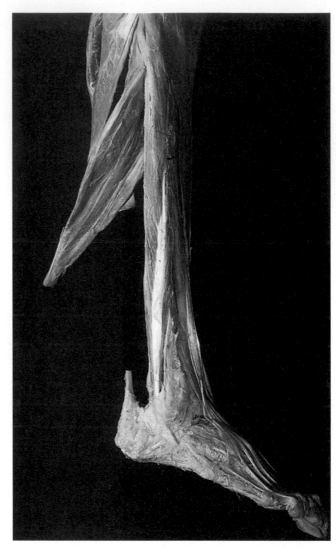

(B)

Figure 5.26   Deep muscles of the lower leg—posterior region.

1. Tibialis anterior muscle
2. Extensor digitorum longus muscle
3. Extensor digitorum brevis muscle
4. Calcaneal tendon (cut)

5. Peroneus longus muscle
6. Soleus muscle
7. Gastrocnemius muscle (lateral head)
8. Peroneus tertius muscle

CHAPTER SIX

Nervous System

The central nervous system (CNS) consists of the brain and spinal cord. It is well protected by bones of the cranium in the case of the brain, and by a bony vertebral column that surrounds and protects the spinal cord. Further, the tough underlying meninges that cover both the brain and the spinal cord consist of three layers: the outer dura mater, the middle arachnoid, and the inner pia mater. Cerebrospinal fluid circulates in the subarachnoid space and cushions the CNS from injury and shock.

The brain and the spinal cord are derivatives of the germinal ectoderm, which initially forms an embryonic neural tube. The neural tube differentiates into the brain and the spinal cord. The brain grows around fluid-filled internal cavities called ventricles. The neural tube in the cephalic region initially differentiates into a forebrain, the prosencephalon; a midbrain, the mesencephalon; and a posterior hindbrain, or rhombencephalon. As the brain grows, the forebrain divides into telencephalon anteriorly, and diencephalon posteriorly. The hindbrain, or rhombencephalon, divides partially into metencephalon and myelencephalon.

As stated earlier, a fully developed brain is housed in a bony cranial vault. In an adult brain the following areas can be identified on a gross level: the cerebrum or cerebral cortex; the diencephalon, which develops into the thalamus, hypothalamus, and the pineal gland; the mesencephalon, which becomes the midbrain; the metencephalon, which becomes the pons and cerebellum; and the myelencephalon, which develops into the medulla oblongata.

The distal end of the brainstem forms a communicating bridge between the brain and the spinal cord. The brainstem continues on into the vertebral column as the spinal cord.

Connecting the brainstem are ten pairs of cranial nerves. The midbrain, or mesencephalon, contains nuclei for cranial nerve III (oculomotor) and cranial nerve IV (trochlear). Also present in the midbrain are the colliculi (two superior and two inferior), and ascending and descending tracks.

The pons and medulla form the lower part of the brainstem. The pons acts as a bridge for ascending and descending fibers between the brain, the spinal cord, and the cerebellum. Nuclei for cranial nerve V (trigeminal), VI (abducens), VII (facial), and VIII (vestibulocochlear), and nuclei for the pontine sleep center and respiratory center are also located in the pons.

The medulla oblongata (medulla) lies inferior to the pons. The medulla displays two prominent enlargements, the pyramids, on its inferior surface. The fibers from the pyramids control skeletal muscle coordination. Also present in the medulla are neurons for the rhythmic cycle of respiration (nuclei for the inspiratory and expiratory centers), and neu-

rons for cranial nerve XI (accessory), and cranial nerve XII (hypoglossal).

The reticular formation is composed of a group of nuclei and their fibers present in the brainstem. The reticular activating system, which constitutes reticular formation and its connections, is a regulatory system that controls the sleep and wake cycle. Acoustic, visual, and mental stimuli can activate the reticular system; conversely, lack of such stimuli can have an opposite effect that leads to inactivity, drowsiness, and excessive sleep patterns. The diencephalon is a derivative of the posterior portion of the forebrain and includes the thalamus, subthalamus, hypothalamus, epithalamus, and their nuclei, and the optic nerves, optic chiasma, the infundibulum, posterior pituitary gland, pineal gland, and mamillary bodies. The two hemispheres of the cerebrum that lie above the diencephalon are separated by a tough dura mater, the falx cerebri. The hemispheres are connected to each other by decussating nerve fibers forming the corpus callosum.

The cerebrum is divided into connecting lobes. The lobes are named after the skull bones that lie above them; thus the lobes are named frontal lobe, parietal lobe, temporal lobe, and occipital lobe. The surface of the lobes is marked by convolutions, the gyri that alternate with shallow depressions or grooves called sulci.

The cerebellum lies inferior to the occipital lobe and posterior to the pons and medulla oblongata of the brainstem. The cerebellum is partially divided into two lateral lobes by a layer of dura mater forming the falx cerebelli. The two lobes of the cerebellum are connected in the middle by a vermis. Each lobe of the cerebellum is essentially composed of an inner white matter of myelinated fibers surrounded by a thin layer of gray matter. The cerebellum is associated with skeletal muscle movements and coordination.

The spinal cord begins at the foramen magnum and terminates between the first and second lumbar vertebrae. The spinal cord is protected by the meninges, the circulating cerebrospinal fluid, and the vertebral column. The spinal cord is partially divided into 31 segments. Each segment of the spinal cord gives rise to a pair of spinal nerves. The spinal nerves branch off into smaller nerves before entering various organs. Cervical enlargement of the spinal cord in the neck region and lumbar enlargements in the lower back region are two prominent areas of the spinal cord, since spinal nerves from these enlargements are part of cervical and sacral plexuses. As the spinal cord descends toward the lumbar region, it narrows into a conical structure, forming the conus medullaris. Extending from the conus medullaris downward is the conical connective tissue sac, which is a continuation of the meninges called the filum terminale.

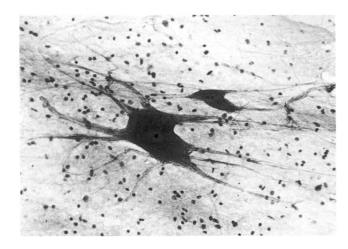

Figure 6.1   Light micrograph (LM) of multipolar neurons. (400×)

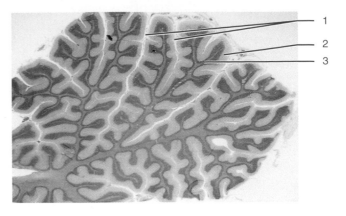

Figure 6.3   LM of a sagittal section through the cerebellar cortex. (1×)

1. Fissures   3. Medulla (white matter)
2. Cortex (gray matter)

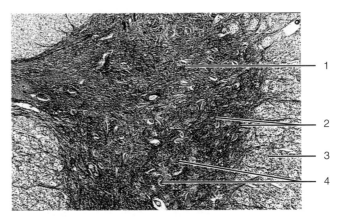

Figure 6.4   LM of gray and white matter as seen in a cross section through the spinal cord. (100×)

1. Gray matter   3. White matter
2. Neuron fibers   4. Neurons

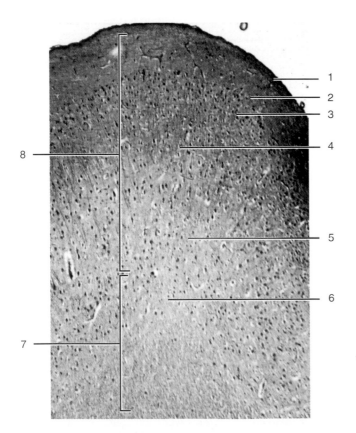

Figure 6.2   LM of a sagittal section through a cerebral cortex gyrus. (40×)

1. Molecular layer   5. Polymorphic cells
2. Pyramidal cells   6. White matter fibers
3. Inner granular layer   7. White matter
4. Inner pyramidal cells 8. Gray matter

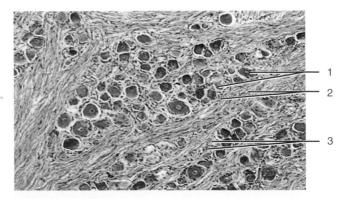

Figure 6.5   LM of a section through a spinal ganglion. (200×)

1. Cell bodies of neurons   3. Satellite cells
2. Nerve fibers

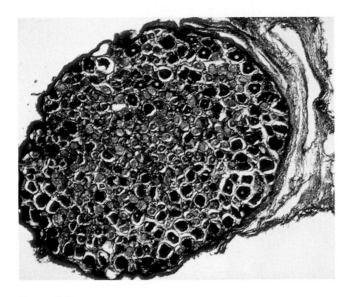

Figure 6.6 LM of a silver-stained section through a peripheral nerve. Myelinated axons can be seen in the fascicles. (100×)

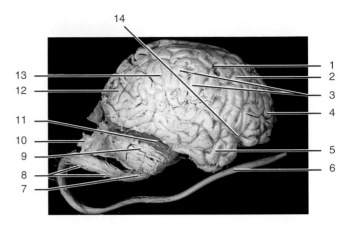

Figure 6.8 The brain and the spinal cord—lateral view.

1. Sulcus
2. Gyrus
3. Parietal lobe
4. Frontal lobe
5. Temporal lobe
6. Spinal cord
7. Medulla oblongata
8. Spinal nerves
9. Cerebellum
10. Dura mater
11. Horizontal fissure of cerebellum
12. Occipital lobe
13. Pia mater
14. Lateral sulcus of Sylvius

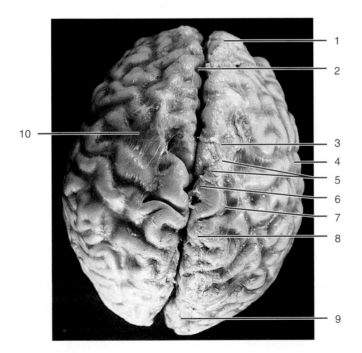

Figure 6.7 Superior aspect of the brain. Right hemisphere of the brain is covered by pia mater and the arachnoid membranes.

1. Frontal lobe
2. Longitudinal fissure
3. Parietal lobe
4. Temporal lobe
5. Arachnoid granulation
6. Precentral gyrus
7. Central sulcus
8. Postcentral gyrus
9. Occipital lobe
10. Precentral sulcus

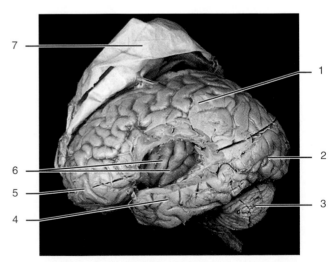

Figure 6.9 Dissection of insula (island of Reil). The insula lies under the temporal lobe of the brain.

1. Parietal lobe
2. Occipital lobe
3. Cerebellum
4. Temporal lobe
5. Frontal lobe
6. Insula (island of Reil)
7. Dura mater, reflected

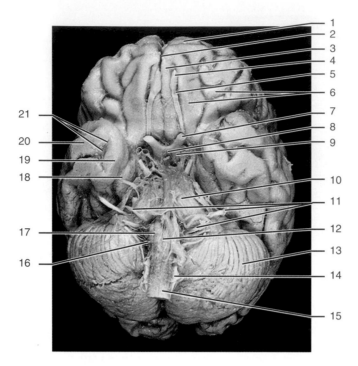

Figure 6.10 Inferior aspect of the brain, brainstem, and cerebellum.

1. Frontal lobe
2. Longitudinal cerebral fissure
3. Gyrus rectus
4. Olfactory lobe
5. Olfactory tract
6. Orbital gyri
7. Optic nerve
8. Optic chiasma
9. Optic tract
10. Oblique fissures of pons
11. Cranial nerves
12. Pyramidal decussation
13. Cerebellum
14. Cervical nerve (C2)
15. Medulla oblongata
16. Cervical nerve (C1)
17. Pyramid
18. Parahippocampal gyrus
19. Occipitotemporal gyrus
20. Inferior temporal gyrus
21. Rhinal sulcus

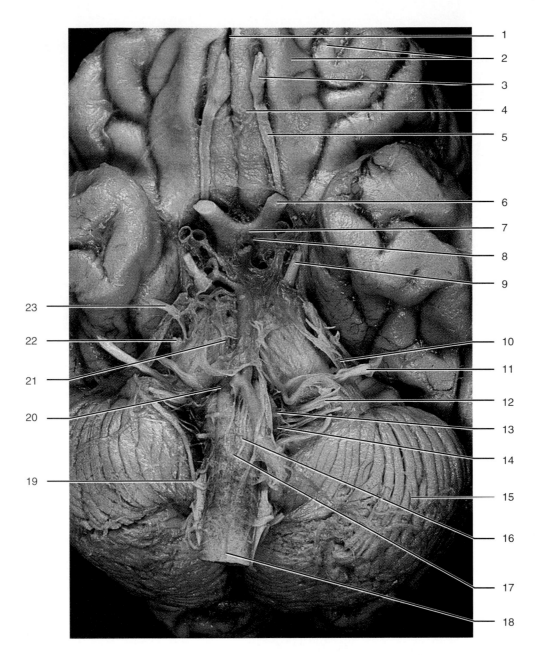

Figure 6.11   Inferior aspect of the brain displaying origin of cranial nerves.

1. Longitudinal fissure
2. Orbital gyri
3. Olfactory bulb
4. Gyrus rectus
5. Olfactory tract (I)
6. Optic nerve (II)
7. Optic chiasma
8. Infundibulum
9. Oculomotor nerve (III)
10. Facial nerve (VII)
11. Vestibulocochlear nerve (VIII)
12. Hypoglossal nerve (XII)
13. Glossopharyngeal nerve (IX)
14. Vagus nerve (X)
15. Cerebellum
16. Medulla oblongata
17. Pyramidal decussation
18. Spinal cord
19. Accessory nerve (XI)
20. Abducens nerve (VI)
21. Pons
22. Trigeminal nerve (V)
23. Trochlear nerve (IV)

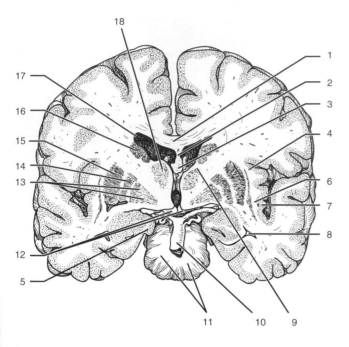

(A)

(B)

Figure 6.12   Middle coronal section of the brain.

1. Corpus callosum (body)
2. Fornix (body)
3. Septum pellucidum
4. External capsule
5. Optic tract
6. Claustrum
7. Insula
8. Lateral ventricle (inferior horn)
9. Internal capsule

10. Basilar artery
11. Pons
12. Mamillary body
13. Globus pallidus internal
14. Globus pallidus external
15. Putamen
16. Caudate nucleus
17. Lateral ventricle (body)
18. Thalamus

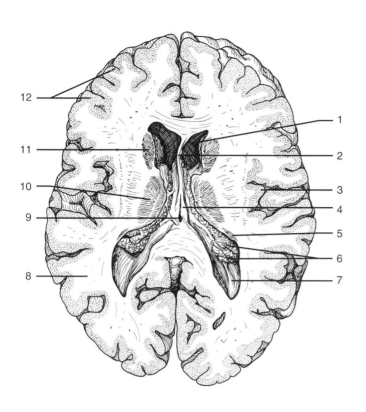

(A)

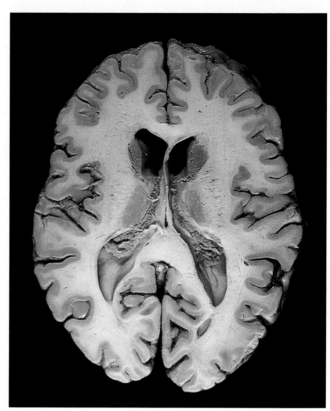

(B)

Figure 6.13 Horizontal section of the brain at the level of the third ventricle.

1. Lateral ventricle (anterior horn)
2. Septum pellucidum
3. Insula
4. Fornix (body)
5. Caudate nucleus (tail portion)
6. Choroid plexus
7. Lateral ventricle (posterior horn)
8. White matter of cerebral medullary center
9. Third ventricle
10. Thalamus
11. Caudate nucleus (head)
12. Cerebral cortex

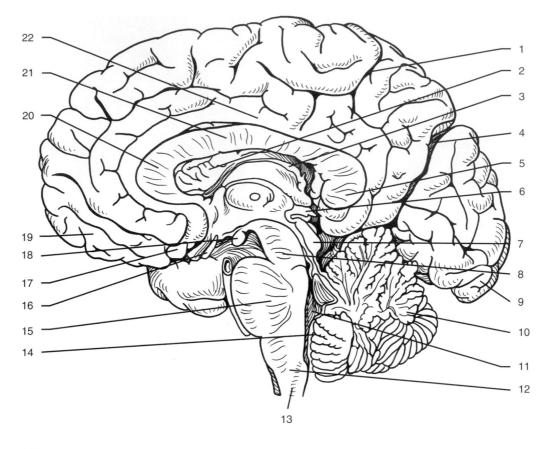

(A)

Figure 6.14   Midline sagittal section of the brain—right side.

1. Central sulcus
2. Lateral ventricle (body)
3. Fornix
4. Parieto-occipital sulcus
5. Splenium of corpus callosum
6. Pineal body
7. Inferior colliculus
8. Cerebral peduncle
9. Occipital lobe
10. Cerebellum
11. Fourth ventricle
12. Medulla oblongata
13. Spinal cord
14. Inferior medullary velum
15. Pons
16. Infundibulum of pituitary
17. Optic chiasma
18. Mamillary body
19. Frontal lobe
20. Genu of corpus callosum
21. Corpus callosum
22. Cingulate gyrus

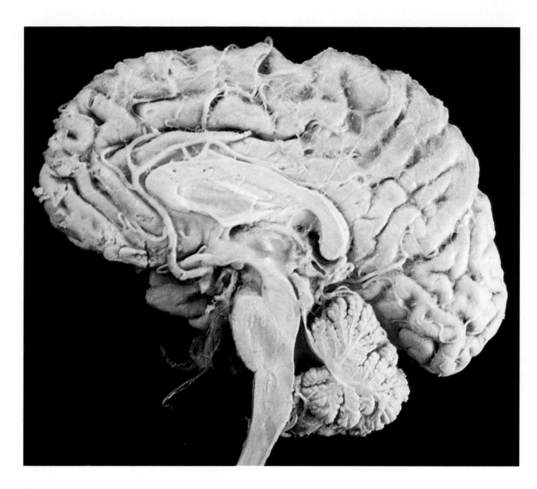

(B)

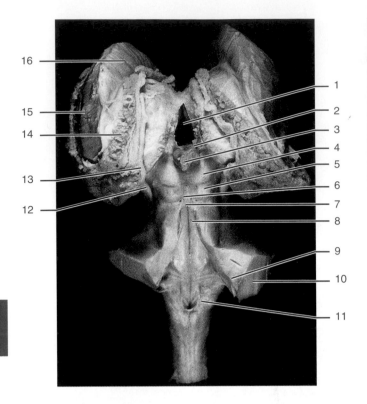

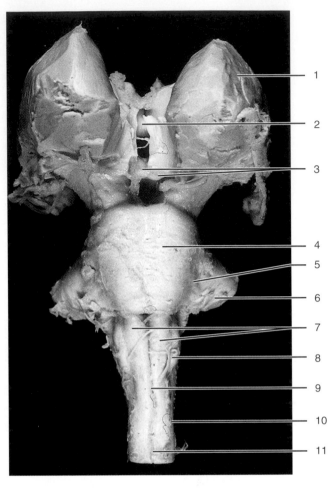

Figure 6.15 Dorsal aspect of the brainstem with cerebellum removed.

1. Third ventricle
2. Habenular trigone
3. Pineal body
4. Superior colliculus
5. Inferior colliculus
6. Frenulum veli
7. Superior medullary velum
8. Facial colliculus
9. Inferior cerebellar peduncle
10. Middle cerebellar peduncle
11. Cuneate tubercle
12. Amygdaloid body
13. Pulvinar of thalamus
14. Choroid plexus of lateral ventricle
15. Internal capsule
16. Head of caudate nucleus

Figure 6.16 Ventral (anterior) aspect of the brainstem.

1. Thalamus (cut surface)
2. Infundibulum
3. Mamillary bodies
4. Pons
5. Cranial nerve
6. Middle cerebellar peduncle
7. Pyramids
8. Olive
9. Pyramidal decussation
10. Spinal nerve (root)
11. Spinal cord

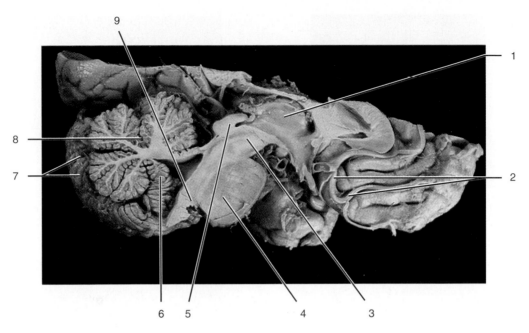

Figure 6.17   Sagittal section of the brainstem, cerebellum, and diencephalon.

1. Thalamus
2. Artery
3. Midbrain
4. Pons
5. Tectum of midbrain

6. Nodules of vermis
7. Cerebellum
8. Primary fissure
9. Medulla oblongata

Figure 6.18 Cerebellum—posterior inferior aspect.

1. Tuber of vermis
2. Pyramid of vermis
3. Cerebellar tonsil
4. Medulla oblongata
5. Pons
6. Central canal
7. Biventral lobule
8. Inferior semilunar lobule

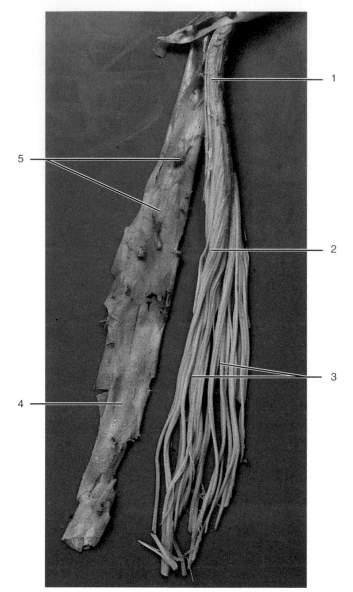

Figure 6.19 Lower end of the spinal cord and cauda equina.

1. Spinal cord
2. Filum terminale
3. Cauda equina
4. Dura mater, lined internally by arachnoid
5. Denticulate ligaments

CHAPTER SEVEN

Special Senses: Olfactory, Gustation, Auditory, Visual

Introduction

Special sensory receptors associated with organs of smell, taste, hearing, equilibrium, and vision are directly or indirectly linked to the nervous system. Information received from these receptors is processed in the brain or the spinal cord. If there is a need for the brain to respond to sensory information, the information goes out from these organs to the effectors via motor or efferent fibers.

Sense of Smell   The olfactory chemoreceptors and epithelial supporting cells are in olfactory organs that are located in the upper regions of the nasal cavity. The olfactory receptors are essentially bipolar neurons scattered throughout the columnar epithelial lining of the mucous membrane. The olfactory neurons at their distal end have knoblike dendrites covered with sensitive cilia that project into the nasal cavity. Olfactory sensations are picked up by these cilia, are passed on to the cell body of the neurons, and from there are sent to the central nervous system via axons.

Sense of Taste   Taste, or gustatory, receptors are concentrated in special organs, the taste buds, which are located in raised dermal papillae of the tongue. Extending from the taste pore of the taste buds are tiny taste hairs (microvilli) of the gustatory, or taste, cells. Surrounding the taste cells in the taste bud are numerous epithelial supporting cells. The hairs of the taste cells act as receptors for the sensation of taste.

Sense of Hearing   The ear acts as an organ of hearing and can be divided into the external, middle, and internal ear. The external ear includes the auricle, external auditory meatus lined by hair, and openings of ceruminous glands. At the end of the external auditory meatus lies the eardrum.

The middle ear includes an air-filled tympanic cavity. The tympanic cavity separates the outer ear from the internal ear. Associated with the tympanic cavity are three small bony auditory ossicles: the malleus, the incus, and the stapes. The ossicles function in transmitting vibrations from the eardrum to the inner ear.

The inner ear is an intricate system of communicating chambers that are connected to bony tubular canals: the osseous labyrinth and the membranous labyrinth. The membranous labyrinth lies within the osseous labyrinth. In between the labyrinths is a narrow space filled with fluid called perilymph. The membranous labyrinth is filled with endolymph.

Within the labyrinths lie the three semicircular canals and the cochlea. The cochlea and the semicircular canals are connected by a bony chamber called the vestibule. The vestibule contains structures for equilibrium and hearing. The cochlea is divided into three channels: the lower channel, the scala tympani; the middle channel, the scala media, or cochlear duct; and the upper channel, the scala vestibuli. The scala tympani and scala vestibuli are filled with perilymph, whereas the scala media contains endolymph. The organ of Corti lies in the scala media.

Sense of Sight   The eye serves as the receptor site for the sense of sight. The eye is supported in its function by the eyelid, lacrimal apparatus, and extrinsic muscles. The eyelid functions as a protective covering for the eye, and its tarsal glands secrete an oily substance that prevents the eyelids from sticking together.

The lacrimal glands of the lacrimal apparatus secrete tears that keep the eyelids and the surface of the eye lubricated. The extrinsic muscles of the eye—the superior rectus, inferior rectus, medial rectus, lateral rectus, superior oblique, and inferior oblique muscles—facilitate the movement of the eye.

The eyeball is a spherical structure that lies in the bony orbit. The wall of the eyeball has three identifiable layers: the inner retina, the middle uveal layer, and the outer corneoscleral coat. The corneoscleral coat is divided into a white, opaque fibrous coat, called the sclera, and a transparent anterior portion, called the cornea. The middle uveal, or vascular, layer is divided into the choroid, the ciliary body, and the iris. The inner retinal layer is a highly complex structure. Its anterior-most layer is lined by the nonsensitive iridial and sensory layers. The posterior sensitive functional layer is lined by photoreceptor organs for the eye; they extend from the anterior ora serrata to the posterior optic papilla of the optic nerve. The fovea centralis, which is surrounded by the macula lutea, or yellow spot, also lies in this layer.

Sense Organs

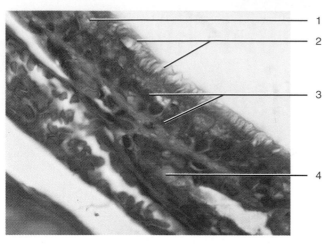

Figure 7.1   Light micrograph (LM) of a section through olfactory epithelium. (400×)

1. Goblet cell
2. Cilia
3. Olfactory and supportive cells
4. Lamina propria

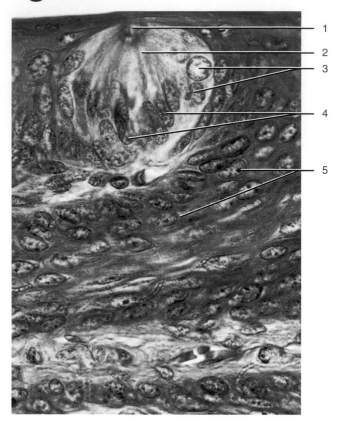

Figure 7.3   LM of a cross section through a taste bud. (1,000×)

1. Taste pore and hair (microvilli)
2. Taste bud
3. Type II cells
4. Type I cells
5. Lamina propria

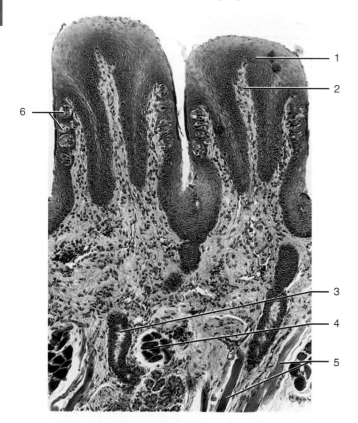

Figure 7.2   LM of a section through circumvallate papillae on the surface of the tongue. (100×)

1. Stratified squamous epithelium
2. Lamina propria
3. Excretory duct
4. Serous gland (of van Ebner)
5. Striated muscle fibers
6. Taste buds

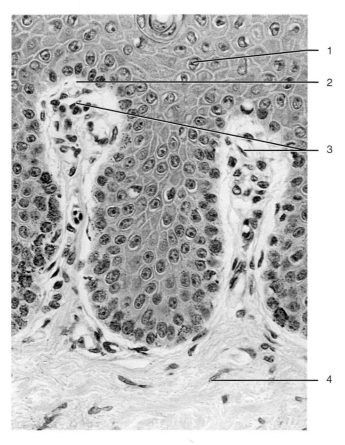

Figure 7.4   LM of a section through tactile Meissner's corpuscles in the dermal papillae of thick skin. (200×)

1. Stratified squamous epithelium
2. Dermal papilla
3. Meissner's corpuscles
4. Lamina propria

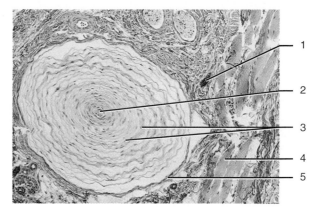

Figure 7.5   LM of a section through a Pacinian corpuscle as seen in the dermis of thick skin. (100×)

1. Blood vessel
2. Bulb of corpuscle
3. Lamellae of corpuscle
4. Dense connective tissue
5. Sheath of corpuscle

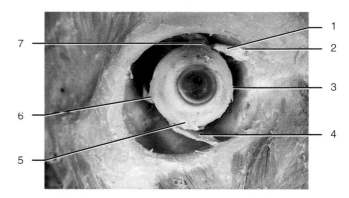

Figure 7.6   Dissection of the left eye, eye orbit, and extraocular muscles.

1. Superior oblique muscle
2. Trochlea
3. Medial rectus muscle
4. Inferior oblique muscle
5. Inferior rectus muscle
6. Lateral rectus muscle
7. Superior rectus muscle

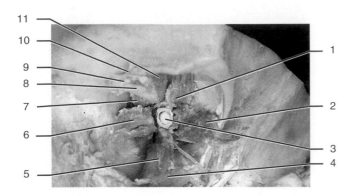

Figure 7.7   Eye orbit and extraocular muscles.

1. Superior rectus muscle
2. Lateral rectus muscle
3. Optic nerve (cranial nerve II)
4. Inferior oblique muscle
5. Inferior rectus muscle
6. Medial rectus muscle
7. Superior oblique muscle
8. Superior oblique tendon
9. Trochlea
10. Supraorbital foramen
11. Levator palpebrae muscle

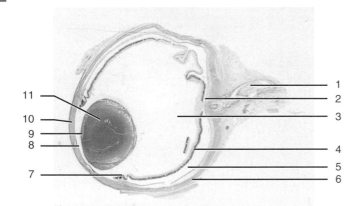

Figure 7.8   LM of a section through a fetal eye. (1×)

1. Optic nerve
2. Optic disc
3. Vitreous body
4. Retina
5. Choroid
6. Sclera
7. Ciliary process
8. Anterior chamber
9. Lens capsule
10. Cornea
11. Lens

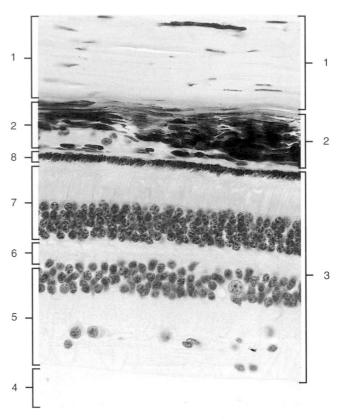

Figure 7.9   LM of a section through sclera, choroid, and retina layers of a fetal eye. (400×)

1. Vascularized sclera
2. Choroid
3. Retina
4. Vitreous body
5. Bipolar neurons, amacrine cells and Müller's cells
6. Outer plexiform layer
7. Rods and cones
8. Pigmented cells of retina

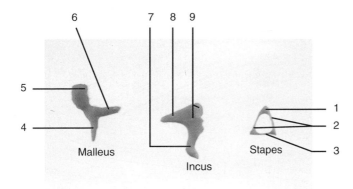

Figure 7.10 Auditory ossicles.

1. Head of stapes
2. Posterior and anterior crura
3. Base of stapes
4. Handle of malleus
5. Head of malleus
6. Lateral process
7. Long crus
8. Short crus
9. Malleus articular facet

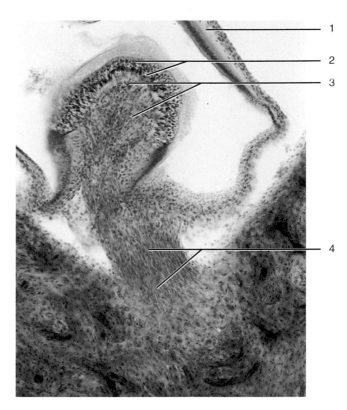

Figure 7.12 LM of a sagittal section through crista ampullaris, which lies in the ampulla of a semicircular canal. (100×)

1. Ampullar membrane
2. Hair cells
3. Sustentacular cells
4. Nerve fibers

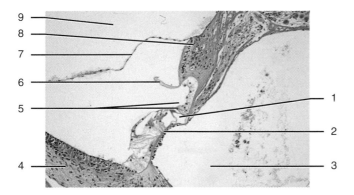

Figure 7.11 LM of a section through cochlear duct and organ of Corti. (400×)

1. Internal spiral nucleus
2. Basilar membrane
3. Scala tympani
4. Osseous lamina
5. Hair cells
6. Tectorial membrane
7. Vestibular membrane
8. Spiral limbus
9. Scala vestibuli

CHAPTER EIGHT

Endocrine System

Introduction

The endocrine system includes special cells, tissues, and organs that function as ductless endocrine glands that secrete their products (hormones) into body fluids such as interstitial fluid, lymph, and blood. The endocrine glands form a separate entity associated with transport of molecules past the cell membrane, control of chemical reactions in cells, and maintenance of water and electrolyte balance. The endocrine glands include the pineal, pituitary, thyroid, parathyroid, and the adrenal glands. However, groups of endocrine cells—the Langerhans cells in the pancreas, Leydig cells in the testis interstitium, and the corpora lutea in the ovaries—are also associated with hormone secretion. Embryologically, the endocrine glands are derivatives of all these germinal layers: the ectoderm, mesoderm, and the endoderm.

Pituitary Gland (Hypophysis)

The pituitary gland is located at the inferior surface of the brain, suspended by a pituitary stalk or infundibulum. The infundibulum is an extension of the hypothalamus and includes the pars tuberalis. The gland lies in a protective bony depression, the sella turcica of the sphenoid bone. Morphologically, the pituitary can be divided into two distinct functional regions or lobes. The anterior pituitary (pars anterior, or pars distalis) secretes human growth hormone (hGH), or somatotropin (STH), a protein; prolactin (PRL), or lactogenic hormone (LTH); gonadotropin, or follicle-stimulating hormone (FSH); thyroid-stimulating hormone (TSH), or thyrotropin; luteinizing hormone (LH); adrenocorticotropic hormone (ACTH); and melanocyte-stimulating hormone (MSH).

The posterior pituitary (pars posterior, or neurohypophysis) consists of neuroglia cells, blood vessels, and nerve fibers extending from paraventricular and supraoptic nuclei located in the hypothalamus. The paraventricular cell bodies secrete oxytocin, whereas the supraoptic nuclei elaborate the antidiuretic hormone (ADH) vasopressin. Both oxytocin and ADH hormones are released by the terminals of the hypothalamic neuron as needed for maintaining homeostasis.

Thyroid Gland

A highly vascular thyroid gland is located inferior to the larynx. The gland is surrounded by a thin connective tissue capsule. The two lobes of the thyroid are connected by an isthmus. Each lobe is partitioned into several lobules by connective tissue. Each lobule is divided into thyroid follicles. The follicular cells of the follicle secrete a jellylike colloidal mixture of thyroglobulin, a combination of glycoprotein and iodated amino acids. Between the follicles lie parafollicular cells (C cells), which secrete calcitonin. Calcitonin and thyroxine are metabolic hormones.

Parathyroid Glands

Two pairs of small, oval-shaped parathyroid glands lie on the posterior surface of the thyroid gland, two on each lobe. Each gland is composed of two types of cells: the chief cells (principal cells) and oxyphil cells. The chief cells secrete a single-chain protein hormone, the parathormone (parathyroid hormone). The parathormone hormone stimulates osteoclast cells, which in turn bring about bone resorption. The function of oxyphil cells is poorly understood.

Adrenal Glands

On each cranial pole of the kidneys lies a pyramid-shaped adrenal gland. The glands are surrounded by a fibrous capsule. In cross section of the gland, two distinct regions, the outer cortex and the inner medulla, can be identified. The hormone-secreting cortex region further differentiates into poorly defined zona glomerulosa, zona fasciculata, and zona reticularis. The zona glomerulosa cells secrete mineralocorticoids (aldosterone and deoxycorticosterone), which regulate electrolyte and water balance. The zona fasciculata cells secrete glucocorticoids (cortisone and hydrocortisone), which increase protein synthesis and decrease stress. The zona reticularis cells secrete male hormone, testosterone; and female hormones, estrogen and progesterone.

Pineal Gland, or Epiphysis Cerebri

The pineal gland is the smallest of all endocrine glands and is located in the third ventricle of the brain. The gland is surrounded by a pia mater capsule. The pineal gland is divided into lobules. Each lobule has a mixture of neuroglia (astrocytes and microglia cells) and pinealocytes (epithelioid cells). In later stages of life the pineal gland goes through retrogressive changes that may lead to increased accumulation of corpora arenacea (brain sand) and connective tissue. Functionally, the pineal gland is associated with secretion of a hormone, melatonin, that may be linked to delayed sexual development, and possibly other substances that may have some metabolic functions.

Endocrine Pancreas

The pancreas is a mixed gland that serves both as an exocrine gland and as an endocrine gland. The exocrine cells of the pancreas secrete digestive enzyme, whereas the endocrine cells, localized as islets of Langerhans, secrete hormones such as glucagon, insulin, and somatostatin, each being secreted by alpha, beta, and delta cells, respectively.

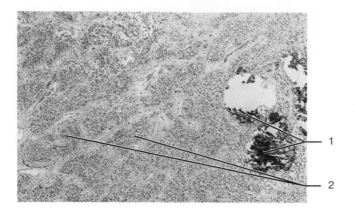

Figure 8.1 Light micrograph (LM) of a section through the pineal gland. (100×)

1. Corpora arenacea (brain sand)
2. Pinealocytes and neuroglia cells

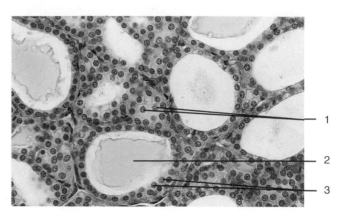

Figure 8.3 LM of a section through the thyroid gland. (400×)

1. Parafollicular cells
2. Follicle with colloid
3. Follicular cells

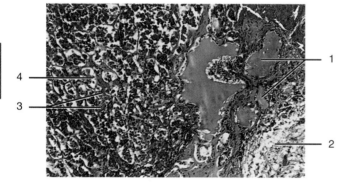

Figure 8.2 LM of a section through the anterior pituitary, pars intermedia, and posterior pituitary. (100×)

1. Colloid
2. Posterior pituitary
3. Blood vessels
4. Anterior pituitary

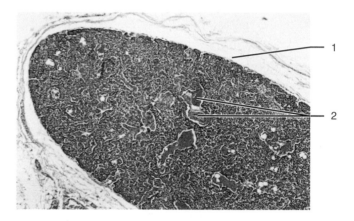

Figure 8.4 LM of a section through the parathyroid gland. (40×)

1. Capsule
2. Trabecular blood vessels

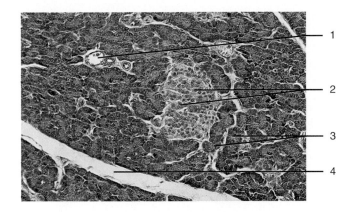

Figure 8.5 LM of a section through an islet of Langerhans in pancreatic tissue. (200×)

1. Blood vessels
2. Islet of Langerhans
3. Pancreatic acini
4. Connective tissue

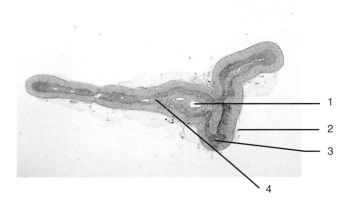

Figure 8.6 LM of a section through the adrenal gland. (1×)

1. Blood vessel
2. Capsule
3. Cortex
4. Medulla

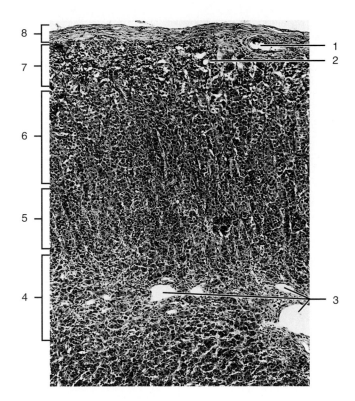

Figure 8.7 LM of a section through the adrenal gland. (100×)

1. Blood vessel
2. Nonmyelinated nerve fibers
3. Medullary veins
4. Medulla
5. Zona reticularis
6. Zona fasciculata
7. Zona glomerulosa
8. Capsule of adrenal gland

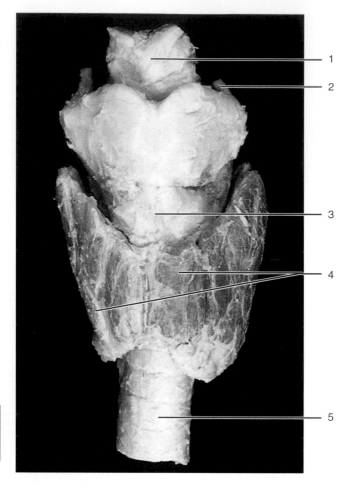

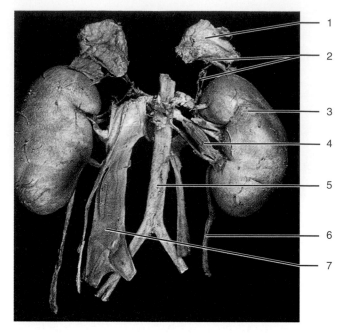

Figure 8.9   Adrenal (suprarenal) glands, kidneys, and blood vessels.

1. Adrenal gland
2. Adrenal artery and vein
3. Kidney
4. Renal artery
5. Abdominal aorta
6. Ureter
7. Inferior vena cava

Figure 8.8   Epiglottis, thyroid cartilage, thyroid gland, and trachea.

1. Epiglottis
2. Greater horn of hyoid bone
3. Thyroid cartilage
4. Thyroid gland
5. Trachea

CHAPTER NINE

Cardiovascular System

Introduction

The function of the cardiovascular system is to circulate the blood, and its components, throughout the body. In the process, the blood delivers nutrients, hormones, and oxygen to the cells, and removes cellular waste and carbon dioxide from the tissues.

The cardiovascular system consists of a muscular heart and associated vasculature. The body wall of blood vessels, except for the capillaries and some smaller blood vessels, are arranged in distinct layers, such as the external tunica adventitia, the middle tunica media, and the inner tunica intima. The preponderance of a given type of tissue in the body wall of a blood vessel depends upon whether the vessel is a vein or an artery.

Arterial Vasculature   On the basis of their morphologies, arteries can be classified as elastic arteries, muscular arteries, and arterioles. Elastic arteries have three well-defined tunics: an outer tunica externa, or adventitia; a prominent media; and an inner intima. The tunica adventitia of elastic arteries is a combination of irregular elastic and collagen fibers concentrically arranged outside the external elastic layer of the tunica media. A well-organized vasa vasorum (blood capillaries) is present in the tunica adventitia.

The tunica media of the elastic arteries is predominantly associated with concentrically arranged elastic fibers. In between fibers lie fibroblasts, smooth muscle, and mesenchyme cells. A small amount of collagen fibers and amorphous ground substance can also be seen in the tunica media.

The tunica intima of the elastic arteries consists of an endothelium that lines the lumen of the blood vessel, elastic fibers scattered in between collagen fibers, and an organized internal elastic lamina infiltrated with amorphous substance.

Muscular Arteries   Muscular arteries also have well-defined tunics. The adventitia is similar to the adventitia of the elastic artery. The tunica media in a muscular artery is highly muscular with concentrically arranged smooth muscle cells. Between muscle cells are a few elastic and collagen fibers. The tunica intima is much narrower, with endothelial cells lining the lumen and an overlying internal elastic layer.

Arterioles   Arterioles are much smaller in diameter than other arteries. The tunics are reduced in their composition, there are fewer elastic and collagen fibers, smooth muscle cells, and endothelial cells in the body wall of the vessel.

Capillaries   Capillaries are the smallest and the simplest form of blood vessels. The body wall is primarily tunica intima with a narrow, overlying tunica media. The tunica adventitia is absent.

Venous Vasculature   The veins have a morphology similar to that of the arteries, but with a thinner body wall and a larger lumen. The veins in general lack external elastic lamina. The internal lamina, if present, is difficult to identify. The veins can be classified as large, medium, small, and venules. The large veins have considerably more collagen than elastic fibers in their body wall. The adventitia in large veins shows concentrations of capillary network, the vasa vasorum. The tunica media is narrow with concentrically arranged smooth muscle, elastic, and collagen fibers. The tunica intima consists of loosely arranged connective tissue and smooth muscle cells. The endothelium lines the lumen. The other veins—medium, small, and venules—retain some of the structure seen in large veins, though at a much decreased level. Collagen fibers are much more common than elastic fibers in the body wall. Few smooth muscle and endothelial cells are retained in the body wall.

The Heart   The heart lies in the left thoracic cavity between the second and fifth rib. It is a highly muscular structure. The body wall of the heart shows three distinct layers: the outer single-layer epicardium, the middle highly muscular myocardium, and an inner endocardium. The thickness of the myocardium varies in different locations of the heart, being the thickest in the body wall of the left ventricle and the interventricular septum.

Also found in the heart is a unique conducting system of Purkinje fibers. The rhythmic action potential for the heart is generated in the sinoatrial and atrioventricular nodes. From these nodes the action potential spreads to all muscle cells of the heart via Purkinje conducting fibers.

Also present in the heart are four chambers with their respective tricuspid, bicuspid (mitral), and semilunar valves of the pulmonary trunk and the aorta.

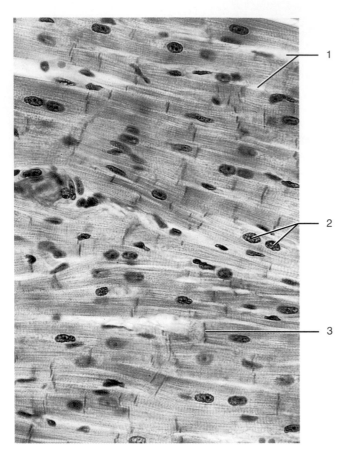

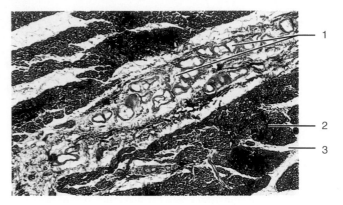

Figure 9.2   LM of Purkinje fibers, which are part of the conducting system of the heart. (200×)

1. Purkinje fibers
2. Regular cardiac muscle fibers
3. Connective tissue

Figure 9.1   Light micrograph (LM) of cardiac muscle in a longitudinal section. (1,000×)

1. Connective tissue
2. Nuclei of cardiac cells
3. Intercalated disc

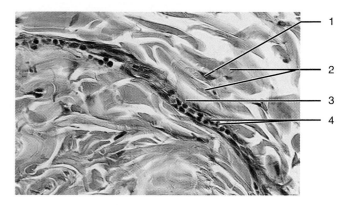

Figure 9.4 LM of an arteriole in a longitudinal section. (400×)

1. Nucleus of a fibroblast
2. Dense collagen fibers
3. Blood vessel
4. Red blood cells in the lumen of the arteriole

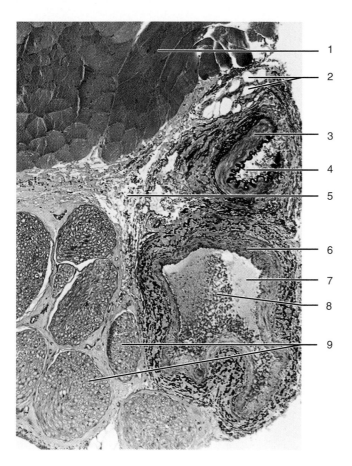

Figure 9.3 LM of a section through a neurovascular bundle. (100×)

1. Skeletal muscle fibers in cross section
2. Adipose cell
3. Body wall of an artery
4. Lumen of the artery
5. Connective tissue
6. Body wall of a vein
7. Lumen of a vein
8. Blood cells and blood plasma
9. Peripheral nerves in cross section

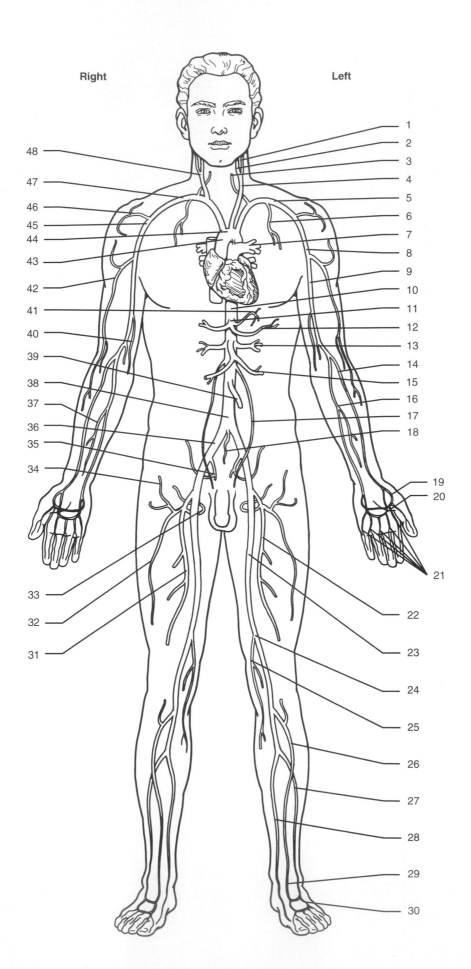

Right

Left

48
47
46
45
44
43
42
41
40
39
38
37
36
35
34

33
32
31

1
2
3
4
5
6
7
8
9
10
11
12
13
14
15
16
17
18

19
20

21

22

23

24

25

26

27

28

29

30

Figure 9.5   Major arteries of the systemic system—anterior view.

1. Left external carotid artery
2. Left internal carotid artery
3. Left vertebral artery
4. Left common carotid artery
5. Left subclavian artery
6. Internal thoracic artery
7. Pulmonary trunk
8. Axillary artery
9. Brachial artery
10. Thoracic aorta
11. Celiac trunk
12. Common hepatic artery
13. Left renal artery
14. Left radial artery
15. Superior mesenteric artery
16. Left ulnar artery
17. Gonadal artery
18. Median sacral artery
19. Deep palmar artery
20. Superficial palmar artery
21. Common palmar digital arch
22. Profundus femoris artery
23. Femoral artery
24. Genicular artery
25. Descending genicular artery
26. Left anterior tibial artery
27. Left peroneal artery
28. Left posterior tibial artery
29. Left dorsalis pedis artery
30. Left dorsal arch
31. Profundus femoris artery
32. Descending branch of lateral circumflex artery
33. Medial circumflex femoral artery
34. Ascending branch of lateral circumflex artery
35. Internal iliac artery
36. Right common iliac artery
37. Anterior and posterior interosseous artery
38. Abdominal aorta
39. Collateral artery
40. Inferior ulnar artery
41. Left gastric artery
42. Profundus brachii artery
43. Ascending aorta
44. Aortic arch
45. Right subclavian artery
46. Descending scapular artery
47. Brachiocephalic (innominate) trunk
48. Right common carotid artery

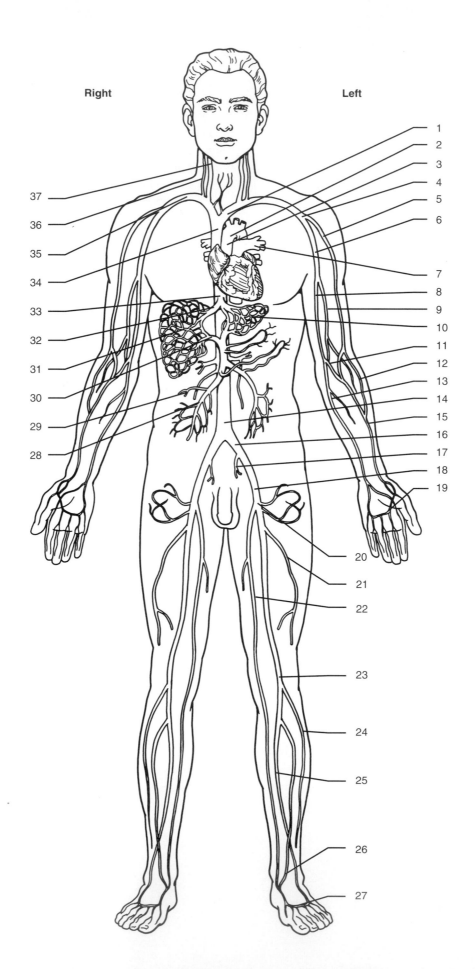

Right

Left

37

36

35

34

33

32

31

30

29

28

1
2
3
4
5
6

7
8
9
10
11
12
13
14
15
16
17
18
19

20

21

22

23

24

25

26

27

Figure 9.6   Major veins—anterior view.

1. Left brachiocephalic vein
2. Aortic arch
3. Ascending aorta
4. Left subclavian vein
5. Left cephalic vein
6. Left basilic vein, upper branch
7. Pulmonary artery
8. Left basilic vein, lower branch
9. Left brachial vein
10. Liver sinuses
11. Left radial vein
12. Left accessory cephalic vein
13. Left median cubital vein
14. Inferior vena cava
15. Left ulnar vein
16. Common iliac vein
17. Internal iliac vein
18. External iliac vein
19. Palmar veins
20. Femoral vein
21. Profundus femoris vein
22. Great saphenous vein
23. Popliteal vein
24. Left short saphenous vein
25. Left anterior tibial vein
26. Left dorsalis pedis vein
27. Dorsal venous arch
28. Renal vein
29. Superior mesenteric vein
30. Hepatic portal vein
31. Liver sinuses
32. Right hepatic vein
33. Inferior vena cava
34. Superior vena cava
35. Right brachiocephalic vein
36. Right external jugular vein
37. Right internal jugular vein

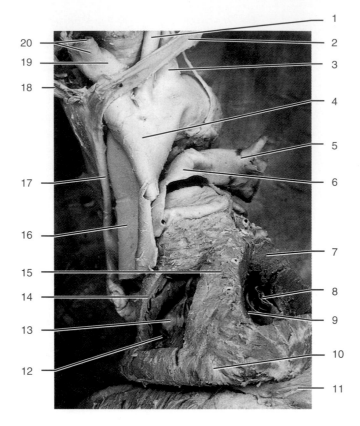

Figure 9.7   Heart, associated blood vessels, and atrioventricular valves—anterior aspect.

1. Left common carotid artery
2. Left brachiocephalic vein
3. Left subclavian artery
4. Aorta
5. Pulmonary artery
6. Pulmonary trunk
7. Body wall of left ventricle
8. Bicuspid (mitral) valve
9. Left ventricle
10. Apex of the heart
11. Diaphragm
12. Right ventricle
13. Tricuspid valve
14. Body wall of right ventricle
15. Interventricular sulcus
16. Ascending aorta, reflected
17. Superior vena cava
18. Right brachiocephalic vein
19. Brachiocephalic artery
20. Right common carotid artery

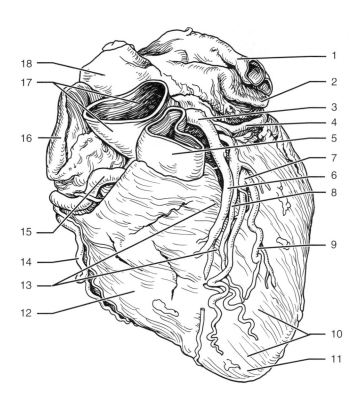

(A)

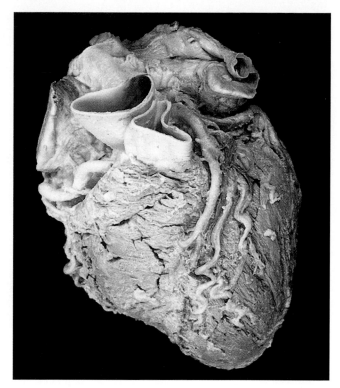

(B)

Figure 9.8   Coronary arteries of the heart—anterior aspect.

1. Pulmonary vein
2. Left auricle
3. Left common coronary artery
4. Circumflex artery
5. Pulmonary trunk
6. Anterior interventricular artery
7. Branch of interventricular artery
8. Great cardiac vein
9. Diagonal branch of left coronary artery

10. Left ventricle
11. Apex of the heart
12. Right ventricle
13. Anterior interventricular sulcus
14. Right marginal artery
15. Right common coronary artery
16. Right auricle
17. Aorta
18. Superior vena cava

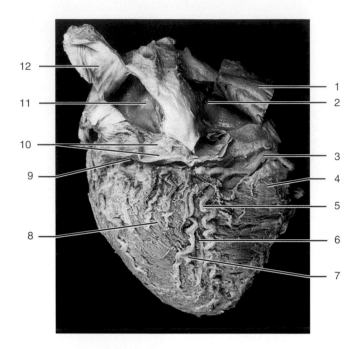

Figure 9.9   Coronary arteries of the heart—posterior view.

1. Pectinate muscles
2. Right atrium
3. Right coronary artery
4. Right ventricle
5. Posterior interventricular artery
6. Middle cardiac vein
7. Posterior interventricular sulcus
8. Left ventricle
9. Atrioventricular sulcus
10. Coronary sinus
11. Left atrium
12. Body wall of left auricle, reflected

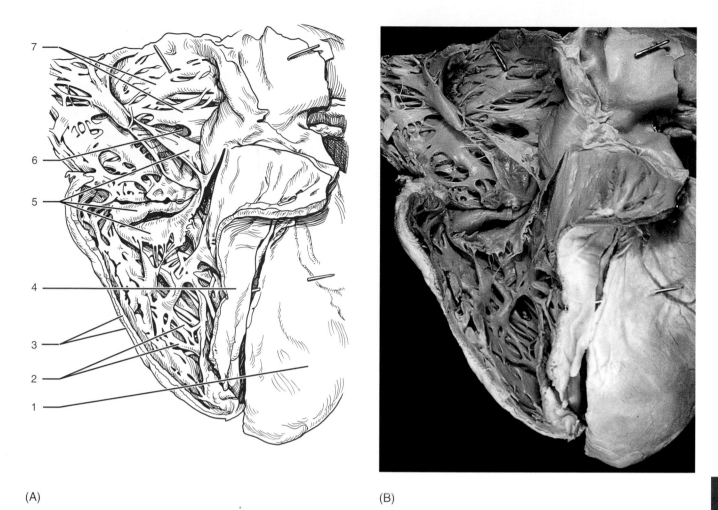

(A)   (B)

Figure 9.10   Right ventricle and right atrial body wall, reflected to show the internal anatomy of the heart.

1. Left ventricle
2. Trabeculae carneae
3. Body wall of right ventricle
4. Interventricular septum

5. Cusps of tricuspid valve
6. Opening of coronary sinus into right atrium
7. Right auricle

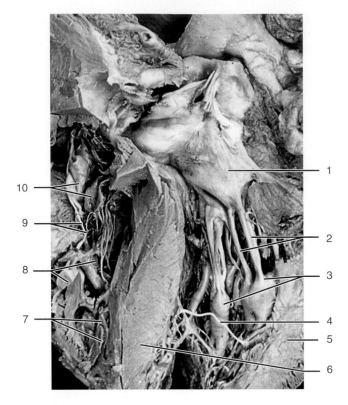

Figure 9.11   Right and left ventricles dissected to show internal anatomy of the heart.

1. Cusps of bicuspid (mitral) valve
2. Chordae tendineae for bicuspid valve
3. Papillary muscle for bicuspid valve
4. Purkinje fiber
5. Body wall of left ventricle
6. Interventricular septum
7. Trabeculae carneae
8. Papillary muscle for tricuspid valve
9. Chordae tendineae for tricuspid valve
10. Tricuspid valve

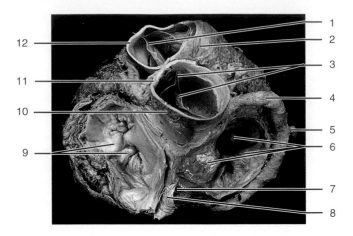

Figure 9.12   Superior aspect of the heart displaying four valves, two atrioventricular, and two semilunar valves.

1. Semilunar valve of the pulmonary trunk
2. Pulmonary trunk
3. Semilunar valve of the aorta
4. Right coronary artery
5. Marginal branch of the right coronary artery
6. Tricuspid valve
7. Coronary sinus
8. Interventricular septum
9. Bicuspid valve
10. Aorta
11. Body wall of the aorta
12. Sinus of pulmonary trunk

CHAPTER TEN

Lymphatic System

Introduction

The immunological defense of the body is associated with the immune or lymphoid system. The organs that are part of the immune system include the tonsils, thymus, lymph nodes, lymph vessels, and the spleen. However, diffused non-capsulated lymphoid aggregates are found in other organs, where they are simply called lymphoid tissue, or nodules.

The lymphatic system is partly responsible for maintaining fluid balance, absorption of lipids and other small molecules from the digestive tract, and producing large numbers of immune-specific cells, the T and B lymphocytes.

Tonsils   Based on their morphology and location, tonsils can be identified as palatine, located at the back of the oral cavity; pharyngeal, found along the posterior wall of the nasopharynx; and lingual, located at the base of the tongue. The lymphoid tissue of the tonsils, in general, forms lymphocyte aggregates, or nodules. The nodules are surrounded by connective tissue forming a loosely arranged capsule. The capsule is covered by stratified squamous epithelium in palatine and lingual tonsils; however, in the pharyngeal tonsil the epithelium is pseudostratified ciliated columnar with interspersed goblet cells.

Thymus   The thymus is surrounded by a thin connective tissue capsule. The gland is divided into two lobes, with each lobe further subdividing into lobules. Each lobule can be identified by a distinct central medulla containing loosely arranged T lymphocytes, and a surrounding cortex with dense concentrations of lymphocytes. Unique to the medulla are Hassall's corpuscles, a concentration of laminated structures of reticular epithelial cells. The function of Hassall's corpuscles is unknown.

Lymph Nodes   Lymph nodes are encapsulated organs found throughout the body, but more so as small aggregates in the cervical, thoracic, and lumbar regions. The nodes function as filtering organs for lymph, sites for division and proliferation of T and B lymphocytes, and storage of lymphocytes in the form of nodules. Morphologically, the node displays an outer cortex region where there are concentrations of lymphoid nodules, and an inner medulla region with densely packed reticulate fibers and histiocytes. In the cortex and medulla, connective tissue trabeculae form channels for the flow of lymph.

Spleen   The spleen is an encapsulated organ that lies in the peritoneal cavity. The splenic capsule is composed of a dense connective tissue lined by a squamous mesothelium. The capsular connective tissue with interwoven reticulate fibers extends deep into the stoma of the spleen to form trabeculae. The splenic parenchyma consists of patches of lymphocytes forming the white pulp. Large concentrations of erythrocytes associated with venous sinuses of the spleen form the red pulp. Reticuloendothelial cells and supporting cells are arranged to form Billroth's, or splenic, cords.

Lymphatic Vessels   Lymphatic vessels have morphologies similar to those of venous blood vessels. They consist of lymphatic trunks, collecting vessels, and blind lymph capillaries. The lymphatic trunks are the largest of the lymph vessels and are similar in morphology to large veins. The tunica intima is lined by endothelial cells. The tunica media has predominantly smooth muscle cells, collagen, and elastic fibers. The tunica externa, or adventitia, consists of collagen and elastic fibers with interspersed smooth muscle cells.

Peyer's Patches   Peyer's patches are concentrations of lymphatic nodules found in the mucosa of the small intestine and the appendix. Secondary nodules may be found in the vicinity of Peyer's patches.

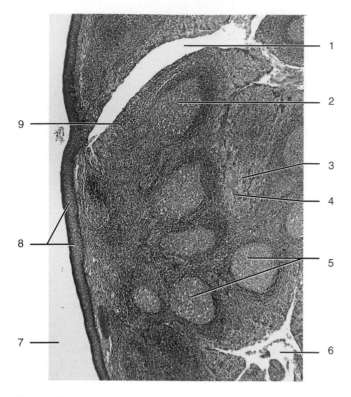

Figure 10.1 Light micrograph (LM) of a section through palatine tonsil. (45×)

1. Crypt
2. Marginal lymph nodule
3. Lamina propria
4. Blood vessel
5. Lymph nodules with germinal centers
6. Connective tissue of the capsule
7. Lumen of pharynx
8. Stratified squamous epithelium
9. Stratified squamous epithelium of crypt

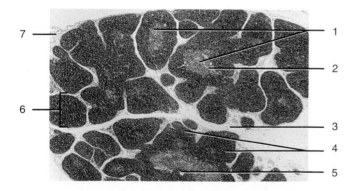

Figure 10.2 LM of thymus gland—a panoramic view. (40×)

1. Medulla of lobule
2. Adipose tissue
3. Trabecula
4. Cortex of lobule
5. Blood vessel
6. Lobule
7. Capsule of thymus

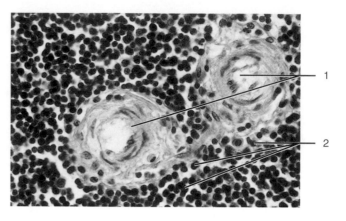

Figure 10.3 LM of Hassall's corpuscles with reticulate epithelial cells. (400×)

1. Hassall's (thymic) corpuscles   2. Lymphocytes

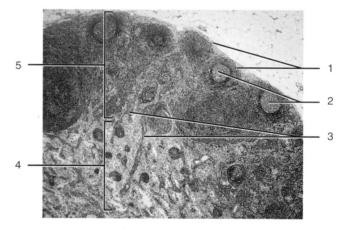

Figure 10.4 LM of a section through a lymph node. (40×)

1. Capsule
2. Germinal center in a nodule
3. Trabeculae
4. Medulla
5. Cortex

Figure 10.5 LM of a lymph vessel with a unidirectional lymph valve. (100×)

1. Valve   2. Endothelium

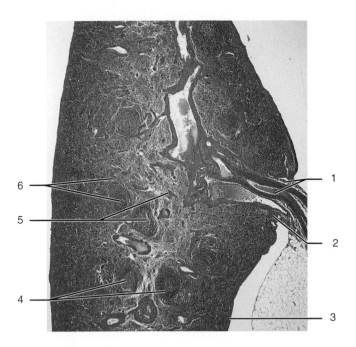

Figure 10.6   LM of a sagittal section through spleen. (20×)

1. Splenic artery and vein
2. Hilum of the spleen
3. Capsule
4. Splenic nodules (white pulp)
5. Trabeculae
6. Venous sinuses (red pulp)

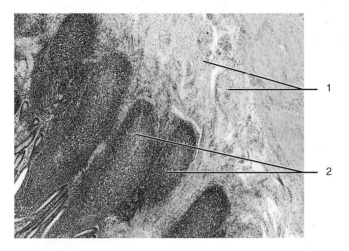

Figure 10.8   LM of Peyer's patches (lymph nodules) as seen in a cross section through the body wall of the small intestine. (40×)

1. Villi of mucosa
2. Peyer's patches (lymphoid nodules)

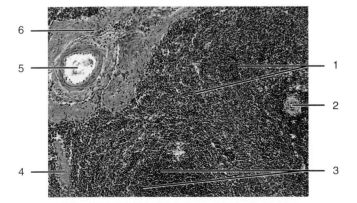

Figure 10.7   LM of splenic tissue displaying red and white pulp. (100×)

1. Red pulp
2. Central artery in splenic nodule
3. White pulp
4. Trabecula
5. Trabecular artery
6. Lamina propria

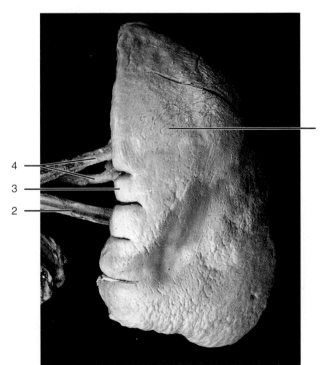

Figure 10.9   Spleen—superior surface.

1. Superior surface of spleen
2. Splenic vein
3. Hilum of spleen
4. Splenic artery

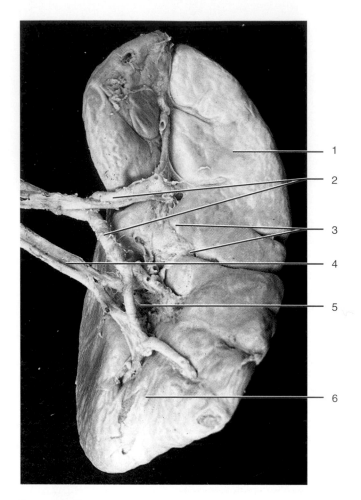

Figure 10.10 Inferior surface of spleen.

1. Gastrosplenic surface
2. Splenic artery
3. Remanents of gastrosplenic ligament
4. Splenic vein
5. Hilum of spleen
6. Colic surface

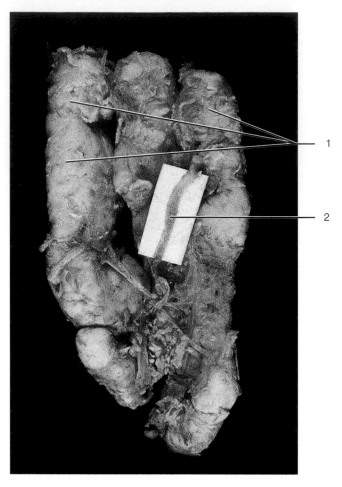

Figure 10.11 A cluster of lymph nodes.

1. Lymph nodes
2. Lymph vessel

CHAPTER ELEVEN
Respiratory System

Introduction

The respiratory system functions as a source of oxygen absorption from external air, while at the same time eliminating carbon dioxide from the blood. Carbon dioxide is a by-product of cellular metabolism. The exchange of gases occurs at two levels: internal respiration, where carbon dioxide diffuses from the cells to the blood, and oxygen from the blood diffuses into the cells; and external respiration, where oxygen from the alveolar air diffuses into the blood, and carbon dioxide from the blood is passed into the alveolar air. Morphologically, the respiratory system is composed of the upper respiratory structures, such as the external nares, vestibule, nasal cavity, turbinate bones (or conchae), paranasal sinuses, the nasopharynx, and the pharynx; and of the lower respiratory tract, which includes the epiglottis, other cartilages within the lungs, bronchioles, alveolar ducts, and alveoli and their surrounding blood capillaries.

The turbinate bones (conchae), the nasal cavity, and the paranasal sinuses (frontal, maxillary, ethmoid, and sphenoid) are covered by a respiratory epithelium supported by a highly vascular underlying lamina propria. The blood vessels, the mucous and serous glands, and the lamina propria connective tissue of the upper respiratory tract form the cavernous erectile tissue. The nasal epithelium is also associated with warming the respiratory air before it enters the trachea and the lungs.

The pharynx, where the air enters from the nasal and oral cavities, is a large cavity antrum. Leading into the pharynx are two openings for nasopharynges, one opening for the oropharynx, one opening for the laryngopharynx, two openings for eustachian tubes, and one opening for the esophagus.

The larynx, a cartilaginous structure, lies between the pharynx and the trachea. The cartilages that form the larynx are the epiglottis, the thyroid, the cricoid, the corniculates, the arytenoids, and the cuneiforms. The cartilages of the larynx, in turn, are connected to thyrohyoid, quadrate, and cricoid membranes. The epithelial lining of the larynx changes according to the location. It is nonkeratinized stratified squamous epithelium in the upper part of the larynx, whereas in the lower region it changes to pseudostratified ciliated columnar epithelium with concentrations of goblet cells.

Trachea

The trachea is a 10–12-cm-long tube with approximately 20 C-shaped cartilage rings that keep the trachea from collapsing. The trachea divides into right and left primary bronchi, which enter the lungs to further divide into secondary and tertiary bronchi in each lung. The tertiary bronchi supply air to 10 respiratory segments in each lung.

The tertiary bronchi subdivide into bronchioles. The bronchioles lack any form of cartilage in their body walls. The body wall comprises an outer fibrous adventitia and an underlying lamina propria with a few smooth muscle cells and elastic fibers. The epithelial lining is of ciliated columnar cells mixed with goblet cells. The bronchioles further divide into terminal and respiratory bronchioles, where the epithelial lining changes to low columnar or cuboidal cells, with an absence of goblet cells.

Alveolar Ducts and Sacs

The alveolar sacs (alveoli) form the terminal structures of the respiratory tree. The narrow alveolar ducts that open into the alveoli are lined by simple squamous epithelium. The alveoli are thin sacs supported by reticular and elastic fibers, simple squamous epithelial cells, the capillary plexus, dust cells (macrophage), septal cells, and a basement lamina.

The blood supply to the lungs is associated with deoxygenated blood going to the lungs by pulmonary arteries, and returning from the lungs, after it has been oxygenated, to the heart via pulmonary veins. Also associated with the lungs are the bronchial arterial system and the azygos venous system.

Figure 11.1   Light micrograph (LM) of a section through nasal mucosa. (1,000×)

1. Cilia
2. Goblet cells
3. Pseudostratified ciliated columnar epithelium
4. Lamina propria
5. Fibroblast

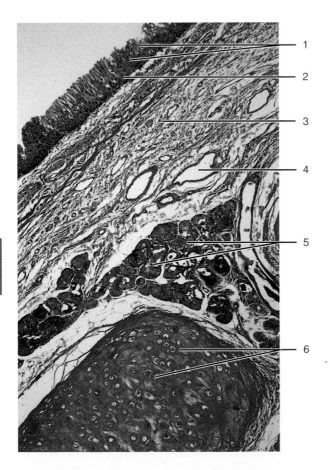

Figure 11.2   LM of a section through larynx. (200×)

1. Cilia
2. Pseudostratified ciliated columnar epithelium
3. Lamina propria
4. Blood vessel
5. Mixed glands
6. Vocal muscle

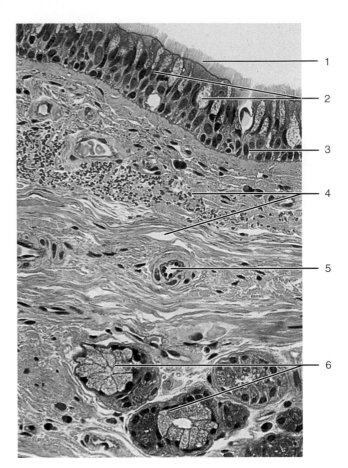

Figure 11.3   LM of a section through body wall of trachea. Hyaline cartilage is not included in this section. (400×)

1. Cilia
2. Goblet cells
3. Pseudostratified ciliated columnar epithelium
4. Lamina propria
5. Blood vessel
6. Tracheal gland

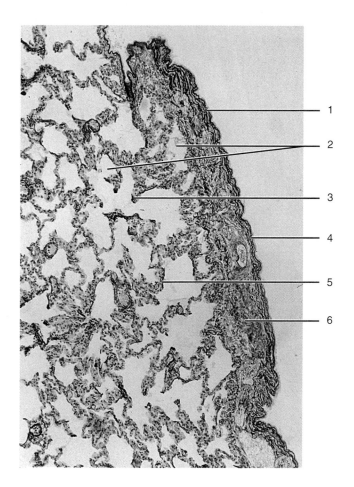

Figure 11.4   LM of lung tissue. (100×)

1. Elastic fibers in visceral pleura
2. Alveoli
3. Body wall of alveolus
4. Visceral pleura
5. Septum
6. Fibrous tissue with elastic and collagen fibers

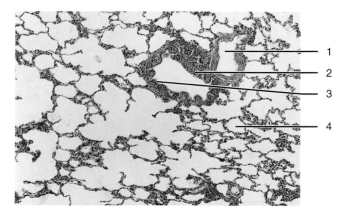

Figure 11.5   LM of a section through lung tissue. Within the lung tissue a bronchiole and a blood vessel can be identified. (100×)

1. Pulmonary artery (branch)
2. Lamina propria
3. Ciliated epithelium
4. Alveolus

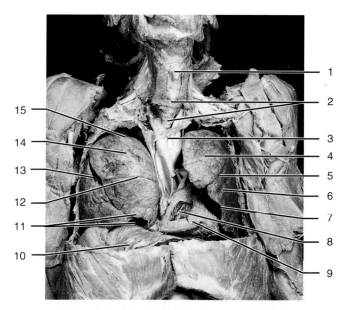

Figure 11.6   Thoracic cavity and viscera.

1. Larynx (thyroid cartilage)
2. Trachea
3. Aorta
4. Left lung (superior lobe)
5. Oblique fissure
6. Inferior lobe
7. Cardiac notch
8. Tricuspid valve of the heart
9. Heart
10. Diaphragm
11. Base of right lung
12. Horizontal fissure
13. Oblique fissure
14. Superior lobe of right lung
15. Apex of right lung

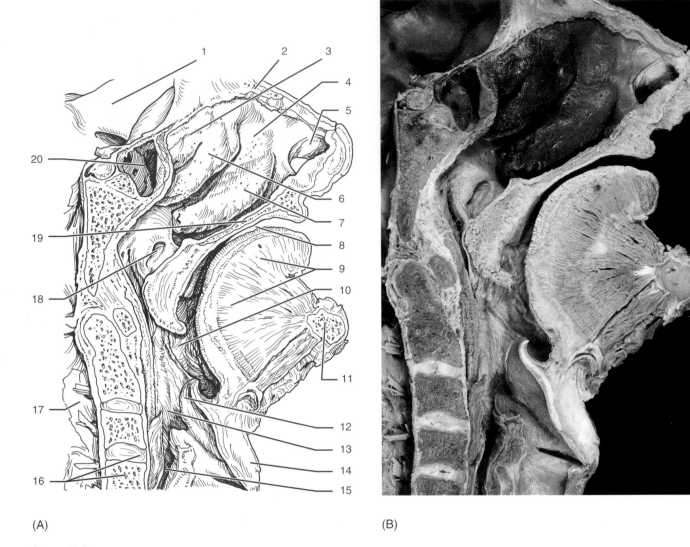

(A)   (B)

Figure 11.7   Medial sagittal section of anterior aspect of the skull to show the respiratory passage.

1. Cranium
2. Nasal bone
3. Superior concha
4. Nasal cavity
5. Vestibule
6. Middle concha
7. Inferior concha
8. Oral cavity
9. Tongue
10. Pharynx

11. Mandible (cut)
12. Epiglottis
13. Pharynx
14. Thyroid cartilage of larynx
15. Esophagus
16. Vertebrae in sagittal section
17. Spinal cord
18. Opening for the eustachian tube
19. Palatine process of maxilla
20. Sphenoidal sinus

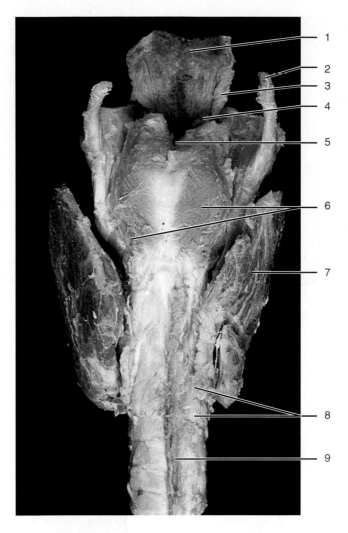

Figure 11.8   Larynx and lobes of the thyroid gland—posterior aspect.

1. Epiglottis
2. Superior cornu of hyoid bone
3. Margin of epiglottis
4. Ventricle
5. Glottis fissure

6. Posterior cricoarytenoid muscle
7. Thyroid gland (lobe—posterior aspect)
8. Tracheal cartilage
9. Trachea

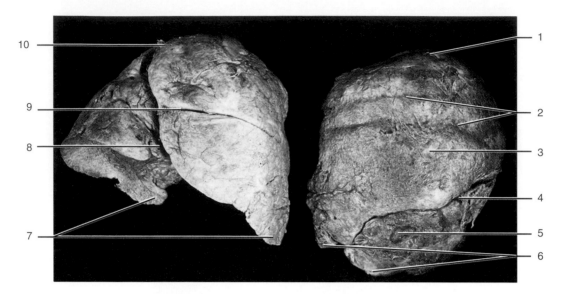

Figure 11.9 Left and right lungs—lateral aspect.

1. Apex of left lung
2. Rib indentations
3. Superior lobe of left lung
4. Oblique fissure
5. Inferior lobe of left lung
6. Base of left lung
7. Base of right lung
8. Oblique fissure of right lung
9. Horizontal fissure
10. Apex of right lung

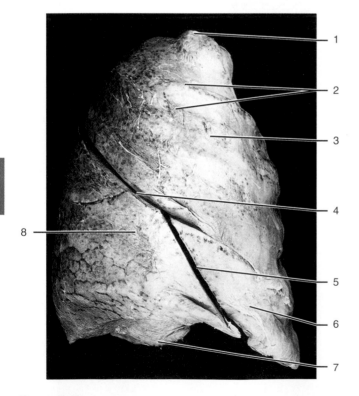

Figure 11.10 Right lung—lateral aspect.

1. Apex of right lung
2. Rib indentations
3. Superior lobe
4. Horizontal fissure
5. Oblique fissure
6. Inferior lobe
7. Base of right lung
8. Middle lobe of right lung

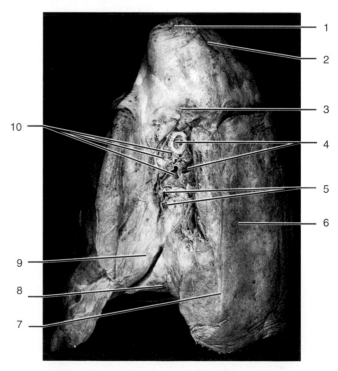

Figure 11.11 Right lung—medial aspect.

1. Apex of right lung
2. Groove for subclavian artery
3. Groove for azygos arch
4. Bronchi
5. Right pulmonary vein
6. Inferior lobe of right lung
7. Pulmonary ligament
8. Diaphragmatic surface
9. Middle lobe
10. Right pulmonary artery branches

CHAPTER TWELVE

Digestive System

Introduction

T he digestive system, or gastrointestinal system, is fairly extensive, beginning at the oral cavity and ending at the anal canal and anus. As the digestive tract courses down from the oral cavity toward the anal canal, it connects to salivary glands, liver, gall bladder, and the pancreas via channels or ducts.

To study the overall morphology of the digestive system, one has to begin with the study of the oral cavity and its structures, then study the esophagus, stomach, small intestine and its divisions, large intestine and its divisions, and finally the morphology of the anal canal and the anus.

The oral cavity is a large cavity with 20 deciduous teeth in children, and 32 permanent teeth in adults. Also present in the oral cavity is a freely movable tongue, and a mucous lining with stratified squamous epithelium. Opening into the oral cavity are salivary ducts associated with the parotid, submaxillary (submandibular), and the sublingual glands. The parotid glands are the largest of the three salivary glands. The parotid glands drain their saliva secretion into the oral cavity by Stenson's ducts, whereas the sublingual glands drain via Rivinus's ducts, and the submaxillary gland drains saliva into the oral cavity by Wharton's ducts.

Tongue
The tongue is a highly muscular structure. The lining on the dorsal surface of the tongue is essentially keratinized stratified squamous epithelium. Embedded in the epithelial lining are three types of papillae: the filiform papillae, fungiform papillae, and 12–15 circumvallate (vallate) papillae. The taste buds are located on the lateral surface of the circumvallate papillae. Below the epithelium of the tongue lies the lamina propria. Deep in the connective tissue of the lamina propria are lingual glands and extrinsic and intrinsic striated muscle fibers. The tongue is attached to the lower palate by a membranous frenulum.

Teeth
A given tooth has an outer enamel covering the crown end of the tooth. Below the enamel lies the dentin, and in the middle of the tooth is the pulp cavity infiltrated by connective tissue. A layer of cementum and a cellular periodontal membrane attach the root of the tooth to the bony sulcus. Also found lining the pulp cavity are odontoblast cells that secrete dentin, and ameloblast cells lining the dentin secrete enamel that forms the crown of the tooth.

Esophagus
The esophagus is a tubular structure that connects part of the pharynx to the cardiac end of the stomach. The mucosal lining of the esophagus is nonkeratinized stratified squamous epithelium. Also present in the body wall is the lamina propria of the mucosa, followed by muscularis mucosae, submucosa, muscularis, with external and internal layers of smooth muscle fibers, and an outer covering of dense connective tissue forming the adventitia.

Stomach
The body wall of the stomach is similar to the body wall of the esophagus, except the stomach mucosa is lined by folds called rugae, and the muscularis has an added internal oblique muscular layer. This layer is not present in any other part of the digestive tract. Embedded in the mucosa layer are gastric pits and gastric glands. The gastric gland cells secrete hydrochloric acid (HCl), proteolytic enzymes, pepsin, mucus, and gastrin-secreting G cells. Gastrin is a hormone that influences the secretion of hydrochloric acid, mucus, and pepsinogen by their respective cells.

Intestinal Tract
The intestine is a hollow, long, and highly convoluted tubular structure that begins with the duodenum at the pyloric end of the stomach and ends at the anus. Similar to the esophagus and stomach, the body wall of the small and the large intestine (colon) also has an outer covering of connective tissue, the serosa. Below the serosa lies the muscularis with external and internal layers of smooth muscle fibers similar to the esophagus, followed by the submucosa and the mucosa. Since the function of the mucosa is absorption in the small intestine (including the duodenum, jejunum, and ileum), the mucosa modifies to form villi lined by goblet cells, microvilli of columnar cells, and undifferentiated columnar cells. The submucosa is highly vascular, has connective tissue, and, within the tissue, large concentrations of mucus-secreting Brunner's glands are common. The muscularis is similar to the muscularis of the esophagus. Common in the mucosa of the small intestine are lymphoid tissue aggregates (nodules) called Peyer's patches.

The large intestine, or colon, is approximately five feet long. On the basis of its arrangement in the abdominal cavity, the large intestine can be subdivided into cecum, ascending colon, transverse colon, descending colon, and sigmoid colon. The morphology of the mucosa changes from that of the small intestine. The villi are absent, and there are large numbers of mucus-secreting goblet cells associated with the mucosa. The muscularis is modified to form three thick muscular bands called taeniae coli. Also associated with the colon are haustra and epiploic appendages.

The rectum is quite similar in its morphology to the colon; however, the crypts of Lieberkühn are few in number and are embedded much deeper in the mucosa.

The anal canal is a continuation of the rectum, the crypts of Lieberkühn are few to almost absent, and the mucosa displays rectal folds (columns of Morgagni). At the distal end of the anal canal are anal valves and sinuses.

Liver   The liver is the largest organ of the body, located just below the diaphragm. The liver is partially divided into four lobes: the right lobe, which is the largest, the left lobe, and the quadrate and caudate lobes. Each lobe is divided into hexagonal lobules, with each lobule having its own central canal. Anastomosing hepatic cords lined by hepatocytes that radiate in all directions from the center of the lobules form the blood-filtering sinusoids. The sinusoids (sinuses) are lined by reticulate fibers, which are infiltrated by phagocytic von Kupffer cells. Portal areas in the vicinity of lobules function as a source of blood to the sinusoid via branches of the hepatic portal vein and hepatic artery. At the same time the branches of the bile duct collect bile from hepatocytes, which is channeled from these cells by canaliculi.

Pancreas   The pancreas is retroperitoneal and functions as an exocrine and an endocrine gland. The endocrine aspect of the gland consists of islets of Langerhans cells that secrete three hormones—glucagon, insulin, and somatostatin—vasoactive intestinal peptide (VIP), substance P, motilin, and pancreatic gastrin.

The exocrine part of the pancreas involves secretion of digestive enzymes by tubuloacinar glands. The pancreatic secretion includes digestive enzymes (lipase, amylase, trypsin, chymotrypsin, elastase, carboxypeptidases, A and B deoxyribonucleases, ribonuclease), hydrochloric acid, and acid-neutralizing bicarbonate. The digestive enzymes and bicarbonate from the pancreas are emptied into the duodenum by the pancreatic and common bile duct.

Gall bladder   The gall bladder lies under the right lobe of the liver, and functions as a reservoir for bile. Bile is also concentrated in the gall bladder. A short cystic duct empties the bile into the bile duct.

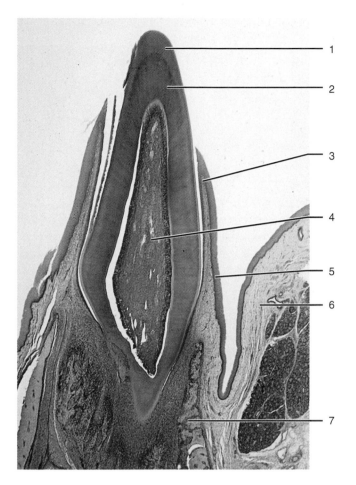

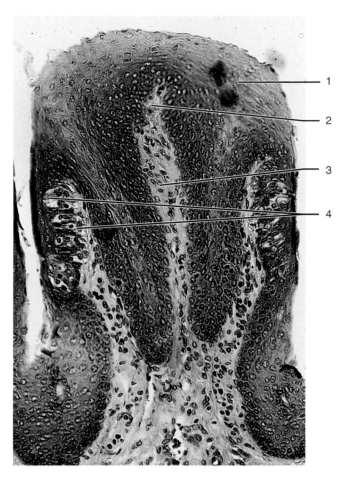

Figure 12.1 Light micrograph (LM) of a sagittal section through a canine tooth and supporting structures. (20×)

1. Enamel
2. Dentin
3. Gingiva
4. Pulp and pulp cavity
5. Stratified squamous epithelium
6. Connective tissue
7. Alveolar bone

Figure 12.2 LM of a section through circumvallate papilla. (200×)

1. Stratified squamous epithelium
2. Dermal papilla
3. Lamina propria
4. Taste buds

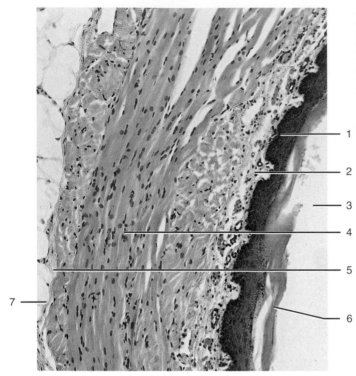

Figure 12.3 LM of a section through the body wall of the esophagus. (40×)

1. Stratified squamous epithelium
2. Lamina propria
3. Lumen of the gut
4. Circular muscle layer
5. Longitudinal muscle layer
6. Mucus
7. Adventitia

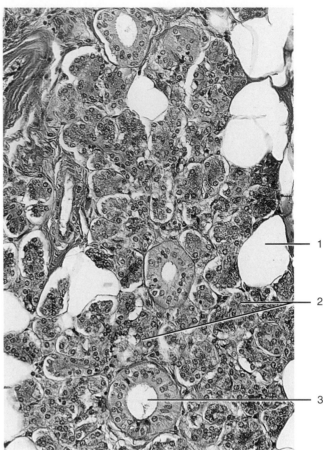

Figure 12.4 LM of a section through a submaxillary gland. (100×)

1. Connective tissue with adipose cells
2. Secretory lobule
3. Excretory duct

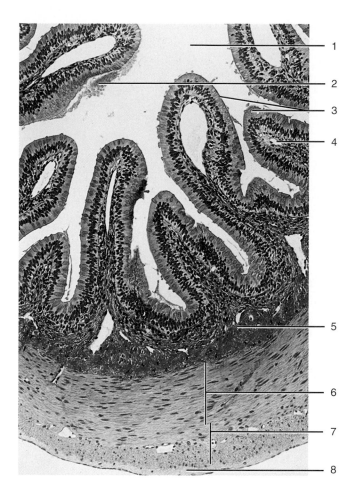

Figure 12.5 LM of a section through the small intestine. (40×)

1. Lumen of the gut
2. Mucus
3. Columnar cells
4. Lamina propria of mucosa
5. Submucosa
6. Circular muscle layer
7. Longitudinal muscle layer
8. Adventitia

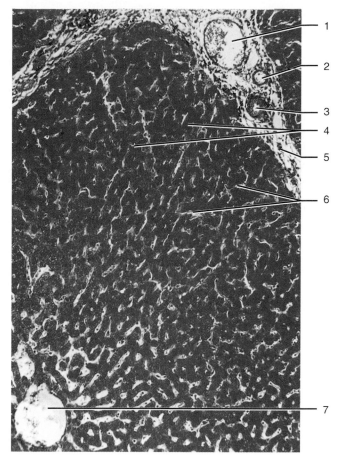

Figure 12.6 LM of a section through liver tissue. (100×)

1. Hepatic portal vein
2. Bile collecting duct
3. Hepatic artery
4. Hepatocytes
5. Connective tissue
6. Liver sinuses
7. Central vein

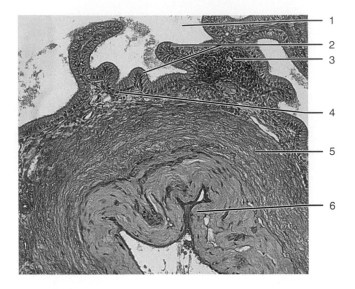

Figure 12.7 LM of a cross section through the gall bladder. (200×)

1. Lumen of gall bladder
2. Columnar cells
3. Mucosa
4. Submucosa
5. Muscle layer
6. Adventitia

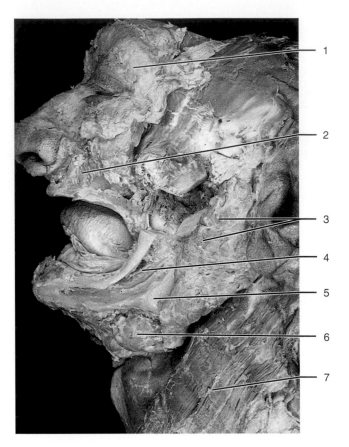

Figure 12.8 Dissection to show the parotid and submandibular glands.

1. Orbicularis oculi muscle
2. Orbicularis oris
3. Parotid gland
4. Buccal branch of facial nerve
5. Mandible
6. Submandibular gland
7. Sternocleidomastoid muscle

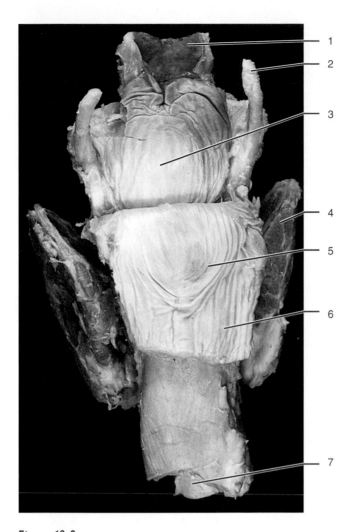

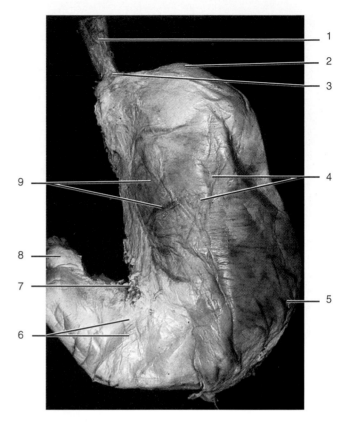

Figure 12.9   Esophagus body wall reflected to show the internal structures.

1. Epiglottis
2. Cornu of hyoid bone
3. Larynx
4. Lobe of thyroid, posterior aspect
5. Esophagus body wall, reflected
6. Rugae of esophagus internal wall
7. Trachea

Figure 12.10   Ventral aspect of the stomach.

1. Esophagus
2. Fundus of stomach
3. Cardiac end of stomach
4. Body of stomach
5. Greater curvature of stomach
6. Pyloric antrum
7. Angular notch of stomach
8. Pyloric sphincter
9. Lesser curvature of stomach

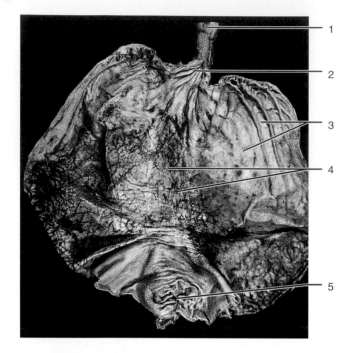

Figure 12.11   Stomach opened to show the internal anatomy.

1. Esophagus   4. Gastric canal
2. Cardiac sphincter 5. Pyloric sphincter
3. Gastric rugae

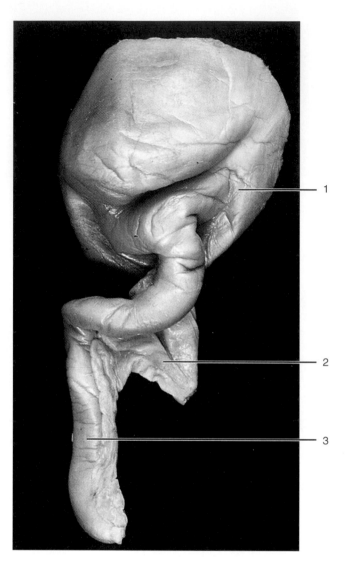

Figure 12.13   Cecum and appendix.

1. Cecum   3. Appendix
2. Mesentery

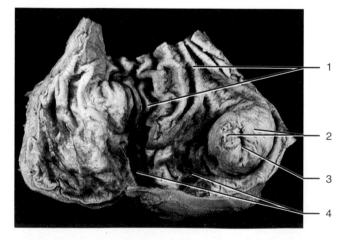

Figure 12.12   Cecum dissected to show the ileocecal valve.

1. Folds of mucous membrane of cecum (rugae)
2. Ileocecal valve
3. Opening of ileocecal valve
4. Antrum of cecum

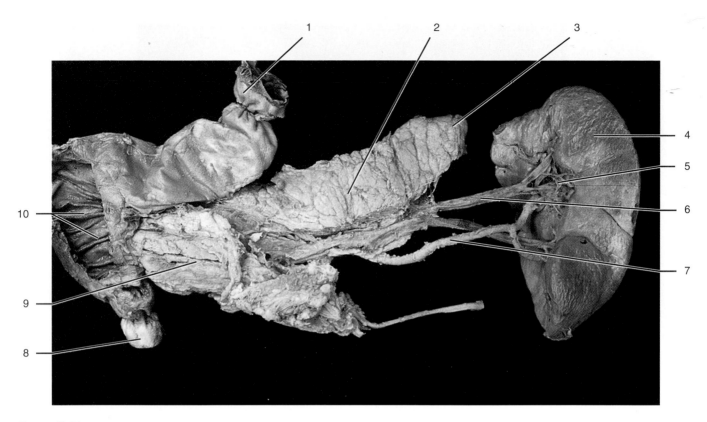

Figure 12.14   Spleen, pancreas, and duodenum.

1. Duodenum, superior end
2. Body of pancreas
3. Tail of pancreas
4. Spleen
5. Hilum of spleen
6. Splenic vein
7. Splenic artery
8. Duodenum, inferior end
9. Head of pancreas
10. Duodenal rugae

Figure 12.15   Liver—superior surface.

1. Left lobe of the liver
2. Falciform ligament of the liver
3. Ligamentum teres

4. Right lobe of the liver
5. Diaphragmatic surface of the liver

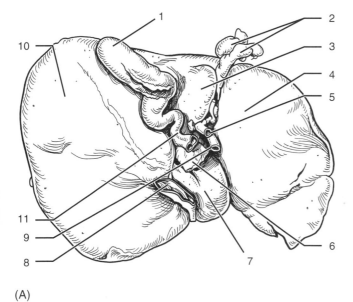

(A)

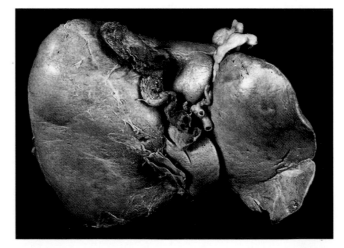

(B)

Figure 12.16   Liver—inferior surface.

1. Gall bladder
2. Ligamentum teres
3. Quadrate lobe of the liver
4. Left lobe of the liver
5. Hepatic artery
6. Hepatic vein

7. Caudate lobe
8. Inferior vena cava
9. Hepatic portal vein
10. Right lobe of liver
11. Cystic duct of gall bladder

CHAPTER THIRTEEN

Urinary System

Introduction

M etabolic waste products from cells that accumulate in the blood plasma, such as electrolytes, nitrogenous compounds, water, and other toxic substances, are systematically eliminated by the urinary system. The filtration of blood within the kidneys is brought about by microscopic filtration units called nephrons. There are approximately one million nephrons per kidney. With a steady supply of arterial blood and blood hydrostatic filtration pressure, the kidneys can produce 180 liters of blood filtrate per day. To maintain a steady filtration pressure, the juxtaglomerular cells of the kidney release renin, a regulatory vasoconstrictor. The kidneys are also responsible for secretory erythropoietin, a hormone that stimulates bone marrow cells to increase erythrocyte production.

Morphologically, in a sagittal section of the kidney, a border area, the cortex, can be differentiated from the underlying medulla, which includes two major calyces and 8 to 12 minor calyces lying in the vicinity of the renal papillae. Ten to 25 collecting ducts open into each papilla. Medullary pyramids with parallel blood vessels and tubules, and renal columns (of Bertin) can also be identified in the medulla of the kidney. Associated with kidney filtration are uriniferous tubules. Each uriniferous tubule is a long tubular structure that can be divided into a nephron and a collecting tubule. The nephrons function in the filtration of blood and formation of filtrate, which is passed on to collecting tubules. The filtrate from collecting tubules is passed on to the calyx minor, from the calyx minor to the calyx major, from there to the pelvis, ureter, and urinary bladder where the filtrate is stored in the form of urine, and finally out of the body upon urination.

Blood Flow through Kidneys   Arterial blood from the abdominal aorta enters the kidney via the renal artery. From there the blood flows to segmental arteries, interlobar arteries, arcuate arteries, interlobular arteries, afferent arterioles, and from the arterioles into the glomerulus where the blood is filtered. After filtration the blood follows a similar path—however, this time through veins. Once the blood is filtered in the Bowman's capsule by glomeruli, the filtered blood enters the efferent arterioles, which in turn drain the blood into the peritubular capillaries and vasa recta. From there the blood continues into the interlobular veins, arcuate veins, interlobar veins, segmental veins, and finally leaves the kidney by the renal vein.

Ureters   The ureters form tubular structures approximately 25–30 centimeters long. They connect the kidney pelvis to the urinary bladder. The ureteral mucosa is lined by transitional epithelium supported by a basement membrane and underlying lamina propria. Below the mucosa is the muscularis consisting of connective tissue and smooth muscle cells.

Urinary Bladder   The filtrate from the kidneys collects in the urinary bladder via the two ureters extending from the kidneys. The body wall morphology of the urinary bladder is similar to that of the ureter, with a few exceptions: the transitional epithelium is thicker, the lamina propria is poorly organized, and the peritoneum covers only the upper surface of the bladder.

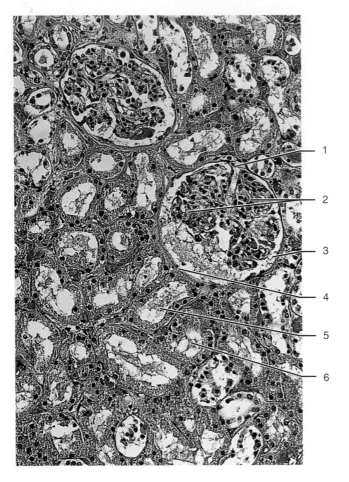

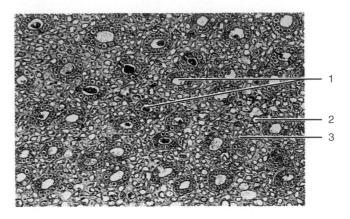

Figure 13.2   LM of a section through the kidney medulla. (100×)

1. Collecting tubules
2. Descending Henle's loop
3. Ascending Henle's loop

Figure 13.1   Light micrograph (LM) of section through the kidney cortex. (200×)

1. Vascular pole
2. Glomerulus
3. Capsular space
4. Bowman's capsule
5. Proximal tubule
6. Distal tubule

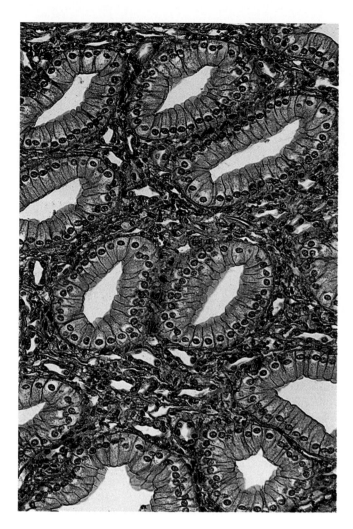

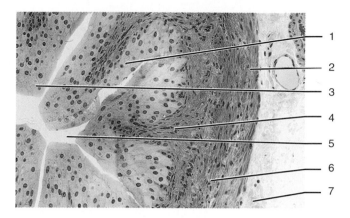

Figure 13.4 LM of a cross section through the ureter. (200×)

1. Transitional epithelium (basal layer)
2. Circular muscle
3. Transitional epithelium (superficial layer)
4. Lamina propria
5. Lumen of the ureter
6. Longitudinal muscle
7. Adventitia

Figure 13.3 LM of a cross section through collecting ducts located in the kidney medulla. (400×)

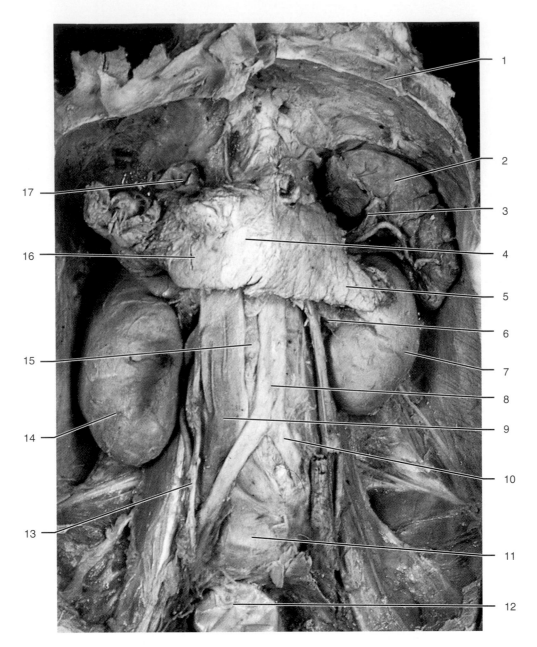

Figure 13.5 Retroperitoneal organs.

1. Diaphragm
2. Spleen
3. Splenic artery
4. Body of pancreas
5. Tail of pancreas
6. Renal artery
7. Left kidney
8. Abdominal aorta
9. Inferior vena cava
10. External iliac artery
11. Rectum
12. Urinary bladder
13. Right ureter
14. Right kidney
15. Psoas muscle
16. Head of pancreas
17. Duodenum

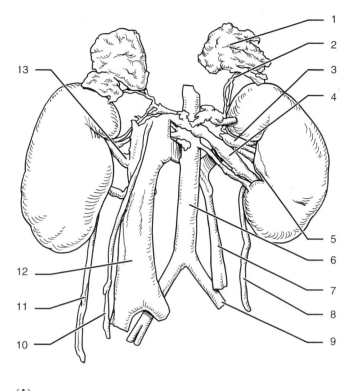

(A)

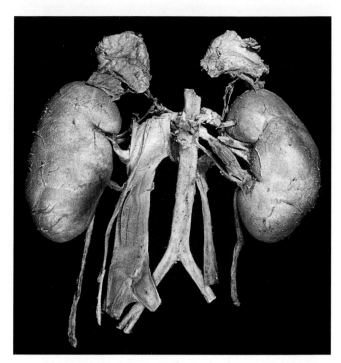

(B)

Figure 13.6   Kidney, adrenal glands, and blood vessels.

1. Suprarenal (adrenal) gland
2. Adrenal artery
3. Renal artery
4. Renal vein
5. Hilum of kidney
6. Abdominal aorta
7. Left gonadal artery

8. Left ureter
9. External iliac artery
10. Right gonadal vein
11. Right ureter
12. Inferior vena cava
13. Right renal vein

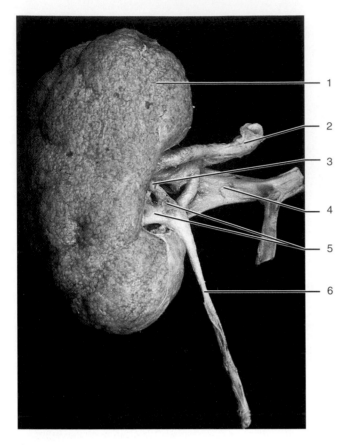

Figure 13.7 Kidney—lateral view.

1. Renal capsule
2. Renal artery
3. Hilum of kidney
4. Renal vein
5. Pelvis
6. Ureter

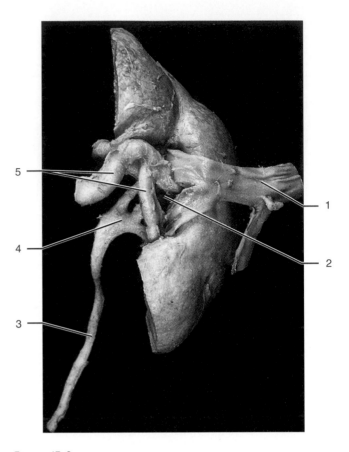

Figure 13.8 Kidney hilum, ureter, and blood vessels—medial view.

1. Renal vein
2. Hilum of kidney
3. Ureter
4. Pelvis of kidney
5. Renal artery

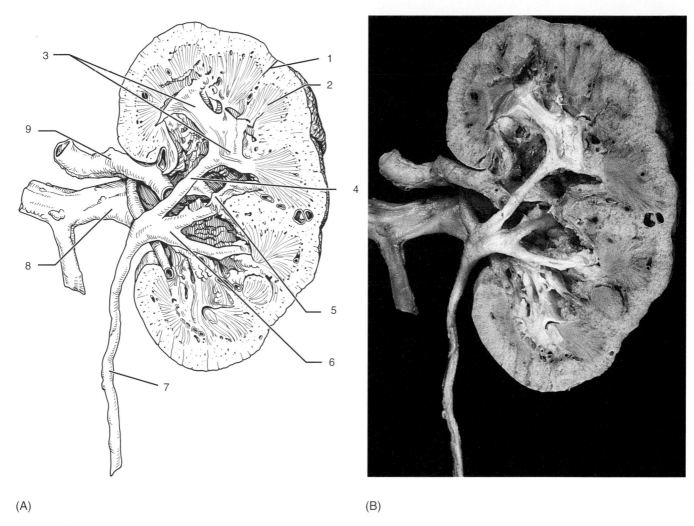

(A)

(B)

Figure 13.9   Midsagittal section of the kidney.

1. Renal cortex
2. Renal pyramid
3. Calyx minor
4. Calyx major
5. Renal medulla

6. Renal pelvis
7. Ureter
8. Renal vein
9. Renal artery

CHAPTER FOURTEEN
Male Reproductive System

Introduction

The male reproductive system primarily consists of organs of spermatogenesis: the testes. Within a testis are seminiferous tubules, the rete testis, the ductus epididymis, the tunics, and the beginning of the ductus deferens (vas deferens). The other supporting structures of the male reproductive system include the rest of the vas deferens, the seminal vesicles, the ejaculatory duct, the corpora cavernosum urethrae, the penile urethra, the penis, and the bulbourethral and the prostate glands.

Testes   The two testes are suspended in the scrotum and surrounded by the tunica vaginalis. Each testis lies in its own dense connective tissue capsule, the tunica albuginea. A highly vascularized connective tissue membrane, the tunica vasculosa surrounds each capsule. Within each testis are approximately 250 pyramid-shaped compartments, the lobuli testis, separated by connective tissue septa. Each lobulus testis may have one to four seminiferous tubules enclosed by a highly vascularized connective tissue. Within the connective tissue that lies between the seminiferous tubules are concentrations of testosterone-secreting interstitial endocrine cells (of Leydig).

The highly convoluted seminiferous tubules are lined by germinal epithelium, which consists of proliferating germinal cells that differentiate into spermatozoa. Supporting cells of Sertoli nourish the spermatozoa to maturity.

Rete Testis   Once the sperm mature, they enter anastomosing channels, the rete testis. From the rete testis channels, the spermatozoa enter the ductuli efferentes. Here the excess fluid secreted by the testis is absorbed. Concentrated spermatozoa from ductuli efferentes are then channeled into the highly convoluted ductus epididymis. The spermatozoa may be retained in the tail of the epididymis for a short period before being passed into the ductus deferens (vas deferens).

Ductus Deferens   The ductus deferens (vas deferens) is a long tubular structure that ascends from the scrotum, courses through the inguinal region, and descends along the lateral wall of the pelvis before merging with the urethra.

The spermatozoa are nourished on their way to the urethra and its opening by seminal vesicle secretion rich in ascorbic acid, prostaglandins, globulin, and fructose, and secretion from the prostate and bulbourethral glands.

Penis   The penis functions as a copulatory organ and a common outlet of urine and seminal fluid, which pass through the penile urethra. The body wall of the penis is composed of three cylinders of erectile tissue: two corpora cavernosa penis, and a single corpus cavernosum urethrae, or corpus spongiosum. The thin, overlying skin of the penis is folded distally to form the prepuce. Proximally, sweat glands mixed with sebaceous glands form the axillary skin of the penis and groin axilla.

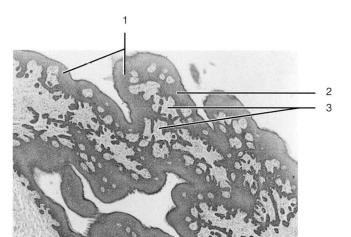

Figure 14.1   Light micrograph (LM) of a section through the prepuce (foreskin) of the penis. (45×)

1. Folds of prepuce
2. Stratified squamous epithelium
3. Connective tissue

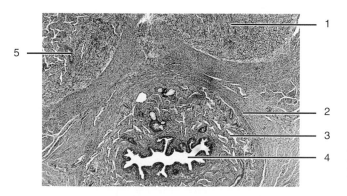

Figure 14.2   LM of a transverse section through the penis. (45×)

1. Corpus cavernosum
2. Tunica albuginea
3. Corpus spongiosum
4. Cavernous urethra
5. Deep artery

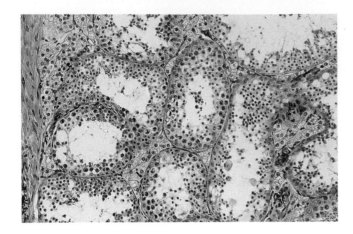

Figure 14.3   LM of testis tissue in cross section displaying seminiferous tubules at low magnification. (200×)

Figure 14.4   LM of seminiferous tubule in cross section at a higher magnification. (400×)

1. Spermatozoa   3. Sertoli cells
2. Lumen of tubule

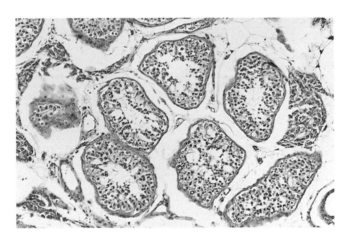

Figure 14.5   LM of rete testis in a cross section displaying rete tubules. Spermatozoa can be seen in the tubules. (100×)

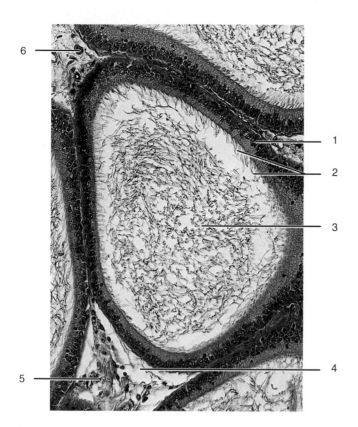

Figure 14.6 LM of a cross section through the epididymis. (200×)

1. Pseudostratified columnar epithelium
2. Stereocilia
3. Spermatozoa in the lumen
4. Lamina propria
5. Smooth muscle fibers
6. Blood vessel

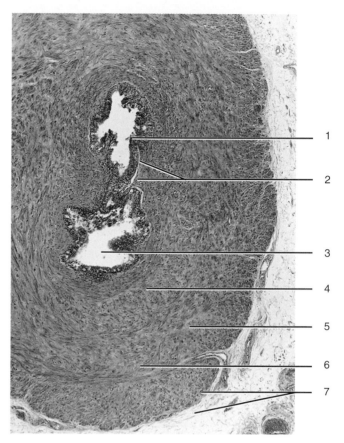

Figure 14.7 LM of a cross section through ductus deferens. (45×)

1. Columnar secretory epithelium
2. Lamina propria
3. Lumen of ductus deferens
4. Inner longitudinal muscle layer
5. Middle circular layer
6. Outer longitudinal muscle layer
7. Adventitia

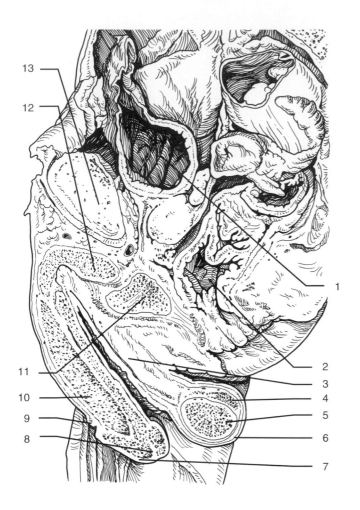

(A)

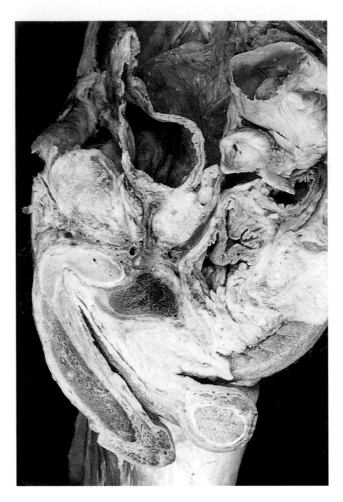

(B)

Figure 14.8 Median sagittal section of male pelvis and perineum.

1. Urinary bladder
2. Colon
3. Ductus deferens
4. Epididymis
5. Seminiferous tubules of testis
6. Scrotum
7. Preputial opening
8. Glans penis
9. Foreskin of prepuce
10. Corpus cavernosum penis
11. Rectum (cut)
12. Prostate gland
13. Symphysis pubis

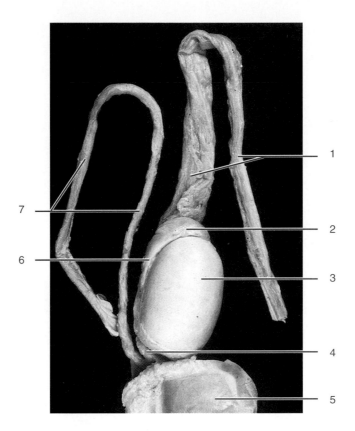

Figure 14.9   Testis and supporting structures.

1. Spermatic cord
2. Head of epididymis
3. Tunica vaginalis (visceral layer)
4. Gubernaculum testis
5. Tunica vaginalis—parietal layer (reflected)
6. Body of epididymis
7. Ductus deferens

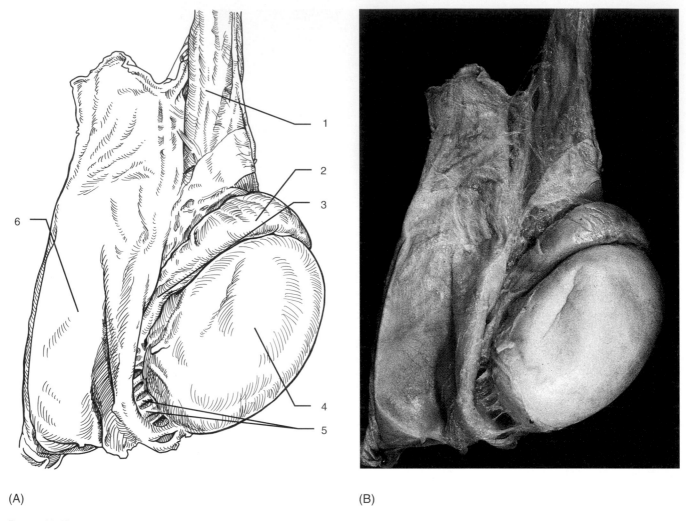

(A)   (B)

Figure 14.10   Testis and supporting structures: spermatic cord and tunica vaginalis.

1. Spermatic cord
2. Head of epididymis
3. Body of epididymis

4. Tunica vaginalis—visceral layer
5. Efferent ductules
6. Tunica vaginalis—parietal layer (reflected)

CHAPTER FIFTEEN

Female Reproductive System

Introduction

The female reproductive organs are specialized organs that produce and sustain a fertilized ovum from a zygote stage to completion and delivery of a fetus. The sex organs of reproduction in females consist of the ovaries, oviducts, uterus, cervix, vagina, and external supporting structures—the labium minus, labium majus, the clitoris, and the milk-secreting mammary glands.

Ovaries The ovaries are two, compact, ovoid structures positioned on either side of the uterus and attached to the body wall by several ligaments. The ovaries function both as an exocrine (cytogenic holocrine) gland producing whole eggs, or ova, and as endocrine glands secreting the sex hormones estrogen, progesterone, and a small amount of testosterone.

Oviducts The two oviducts, or fallopian tubes, transport the ovum from the ovaries to the uterus after it is delivered to a funnel-shaped infundibulum at the distal end of the oviduct.

Uterus The uterus is a pear-shaped, hollow, muscular structure, approximately five to seven inches long. The body wall of the uterus consists of an outer single-layer perimetrium, a thick muscular myometrium in the middle, and an inner lining of endometrium that changes in thickness during the estrus cycle. The endometrium is lined by simple columnar epithelium.

Cervix The cervix opens into the vagina and forms the inferior segment of the uterus. The body wall of the cervix consists of dense collagenous connective tissue lined internally by tall mucus-secreting columnar cells, some of which may be ciliated. Plicae palmatae form deep clefts, or furrows, in the lining of the cervix.

Uterine Lining The lining of the uterus, the endometrium, at puberty and thereafter up to menopause, goes through cyclic changes associated with estrus stages. During an estrus cycle the endometrial changes consist of the proliferation (follicular) stage, and the progestational (luteal) stage; ischemic stages are associated with cessation of blood flow to the blood vessels of the endometrium, and the disintegration of the endometrial lining.

Vagina The vagina encompasses the lowermost part of the female reproductive tract. The internal wall of the vagina consists of transverse folds, or rugae, lined by stratified squamous epithelium. Below the vaginal epithelium lies the connective tissue lamina propria. Smooth muscle bundles are associated with the lamina propria.

Mammary Glands Mammary glands are modified sweat glands lying in the subcutaneous tissue. Each mammary gland is composed of 15–20 lobes surrounded by connective tissue and concentrations of adipose tissue. Each lobe is further divided into many lobules separated by adipose and connective tissue. The secretory product (milk) is collected from each lobule via an intralobular duct. The intralobular ducts join to form the interlobular ducts. The interlobular ducts from each lobe of the gland join to form the lactiferous duct that terminates as the lactiferous sinus just before the opening in the summit of the nipple.

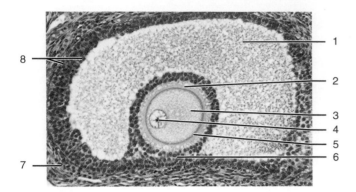

Figure 15.1 Light micrograph (LM) of a section through secondary oocyte. (200×)

1. Antrum
2. Corona radiata
3. Ovum
4. Nucleus
5. Zona pellucida
6. Cumulus oophorus
7. Ovary cortex
8. Theca interna and externa

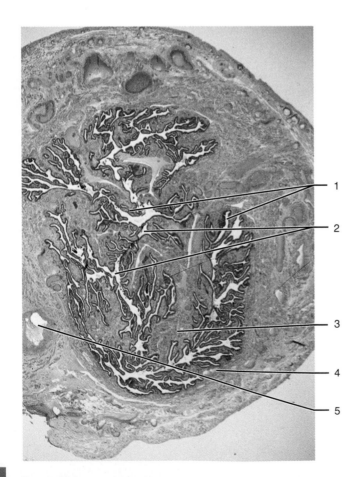

Figure 15.2 LM of a cross section through ampulla of the oviduct. (20×)

1. Ciliated columnar epithelium
2. Lumen of oviduct
3. Lamina propria
4. Circular muscle fiber
5. Blood vessel

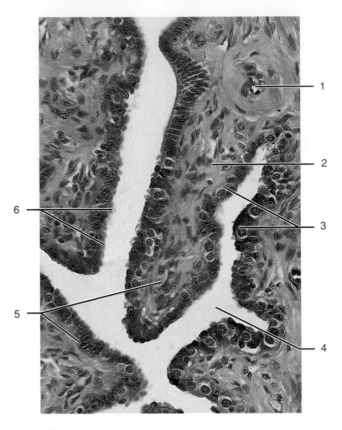

Figure 15.3 LM of mucosal folds of the oviduct. (40×)

1. Blood vessel
2. Lamina propria
3. Ciliated columnar epithelium
4. Lumen of oviduct
5. Mucosal folds
6. Cilia

Figure 15.4 LM of endometrium—proliferative (follicular) stage. (100×)

1. Functionalis
2. Basalis layer
3. Myometrium

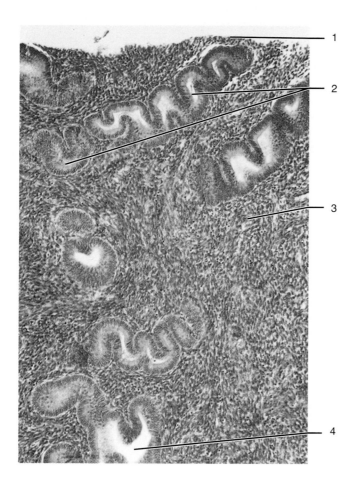

Figure 15.5   LM of endometrium—secretory stage. (100×)

1. Intact columnar epithelium
2. Uterine gland
3. Intergranular lamina propria
4. Dilated uterine gland with secretion

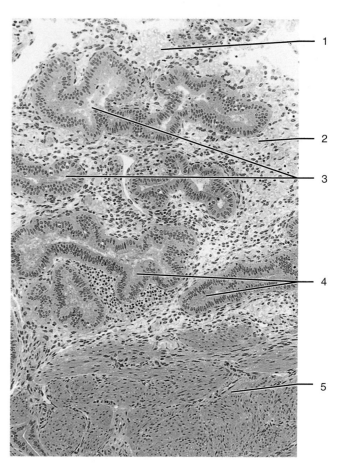

Figure 15.6   LM of endometrium—menstrual stage. (100×)

1. Endometrium with disintegrated epithelium
2. Fragmented stroma
3. Disintegrating glands
4. Glandular lumen filled with blood
5. Smooth muscle fibers of myometrium

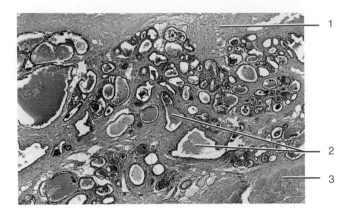

Figure 15.7 LM of a section through lactating mammary gland. (40×)

1. Adipose tissue
2. Active alveoli with secretion
3. Lamina propria

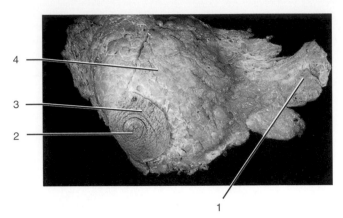

Figure 15.8 Mammary gland.

1. Axillary tail
2. Nipple
3. Areola
4. Mammary gland

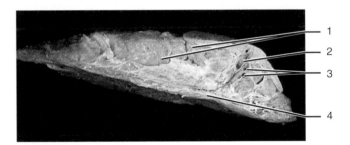

Figure 15.9 Mammary gland—sagittal section.

1. Mammary tissue
2. Lactiferous duct
3. Lactiferous sinus
4. Pectoralis major muscle

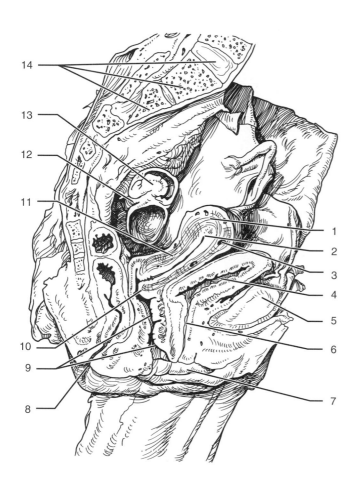

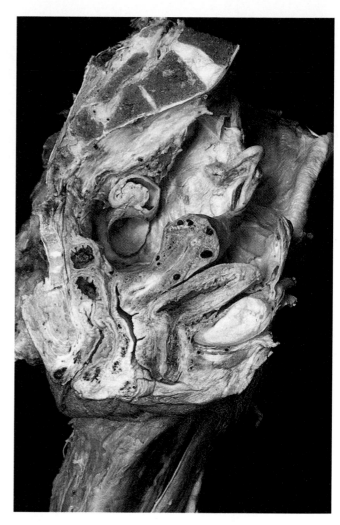

Figure 15.10 Median sagittal section of female pelvis.

1. Myometrium of fundus
2. Endometrium
3. Blood vessel
4. Urinary bladder
5. Symphysis pubis
6. Urethra
7. Labium minus
8. Anus
9. Vagina
10. Cervix
11. Uterus
12. Oviduct
13. Ovary
14. Vertebrae

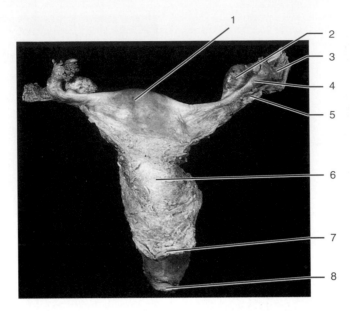

Figure 15.11   Ovaries, uterine tubes, and uterus.

1. Fundus of uterus
2. Ovary
3. Fimbriae of uterine tube
4. Uterine tube

5. Broad ligament
6. Body of uterus
7. Cervix
8. Vaginal wall

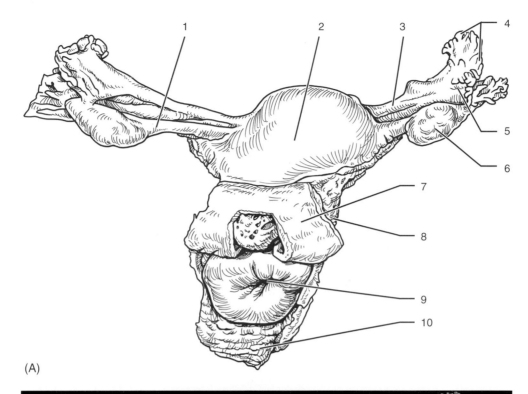

(A)

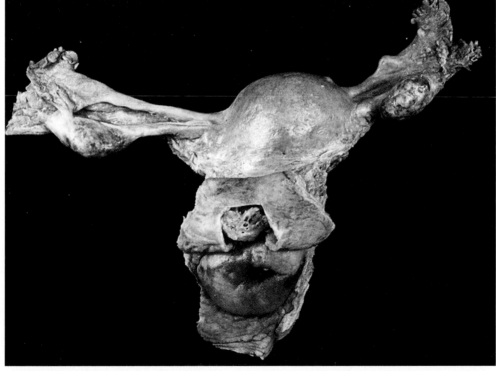

(B)

Figure 15.12   Vaginal wall cut and reflected to expose the cervix.

1. Ovarian ligament
2. Fundus of uterus
3. Uterine tube
4. Fimbriae of uterine tube
5. Infundibulum
6. Ovary
7. Vaginal wall, reflected
8. Body of uterus
9. Cervix of uterus
10. Vaginal wall

INDEX